Patrick Moore's Practical Astronomy Series

For other titles published in the series, go to
www.springer.com/series/3192

Make Time for the Stars

Fitting Astronomy into Your Busy Life

Antony Cooke

 Springer

Antony Cooke
Capistrano Beach, CA

ISBN 978-0-387-89340-2   e-ISBN 978-0-387-89341-9
DOI: 10.1007/978-0-387-89341-9
Springer Dordrecht Heidelberg London New York

Printed on acid-free paper

Springer is part of Springer Science+Business Media (www.springer.com)

This book is dedicated to my mother, Margot Cooke, with love and appreciation for the encouragement, freedom, and opportunities she provided so that I could make the most of everything meaningful to me.

About the Author

It has often been said that music and astronomy go hand in hand. **Antony Cooke's** passion for both fields were clear very early in his life, but music ultimately would claim his career. A cellist of international renown, Cooke has been one of the leading players in the Hollywood recording industry for many years, having been Associate Professor of Cello at Northwestern University in Chicago until 1984. A US citizen but born in Australia and educated in London, he received artist diplomas from the Royal College of Music and the Royal Academy of Music. During this time he was a recipient of numerous prizes and awards, including the Gold Medal at the London Music Festival. Becoming England's youngest principal cellist (London Mozart Players), his career grew as international soloist, solo recording artist, university professor, and published composer, including for prime time television.

The dual nature of Cooke's interests continued, astronomy remaining the counterbalance in his life. Always looking for ways to improve his experience at the eyepiece, he has constructed many telescopes over the years, with increasing sizes being the hallmark of his often quirky designs. With limited time becoming an increasing factor, and with simplicity in mind, Cooke made the conscious choice not to be a slave to the new directions of mainstream amateur astronomy. Blending some of the best that modern technology has to offer with the careful choice of portable equipment, Cooke's somewhat unorthodox approach, developed over the course of his lifetime, has proved to be his best solution.

Make Time for the Stars is Antony Cooke's third book on astronomy, preceded by *Visual Astronomy in the Suburbs* (Springer 2003) and *Visual Astronomy under Dark Skies* (Springer 2005).

Contents

Section I The Busy Astronomer

Chapter 1 Introduction 3

Chapter 2 **Making the Right Equipment Choices** 9
From Absolute Zero to Absolute Minimum 10
Potential Pitfalls – Do Not Fall In 11
The Homebuilt Telescope 16
Eyepieces 20
Right on the Money 21
Aperture Fever 22
Other Worthy Accessories 23

Chapter 3 **How to Expand Your Potential** 25
The Best of the Bunch 26
Comparing CCD Video Cameras and Image Intensifiers 43
A Word on Private Observatories 44
And Finally 45

Chapter 4 **Maximizing Your Time at the Telescope** 47
The Importance of Precise Optical Alignment 47
Quick Setup Project: Collimation Made Easy 49
The Importance of Clean Optics 51
Quick Setup Project: Easy Cleaning of Optical Components 52
Getting More from Your Newtonian 52
Other Distractions 55
The Weather! 57
The Value of True Portability 58

Section II **The Moon**

Chapter 5   The First Port of Call 63
Quick Project: Lunar Fly-By.. 64
A Real Lunar Fly-By!... 65

Chapter 6   The Moon: Close Up and Personal 67
Choosing a Telescope.. 68
Quick Project: Comparing Lunar Features
    to Familiar Landmarks... 69
Flying with *Apollo* 70
Quick Project: Examining Mountainous Contours
    at the Lunar Limb.. 71
Apollo 71
Revisiting Familiar Lunar Features... 73
Quick Project: An *Apollo* Mission Relived.................................... 73

Chapter 7   Instant Imaging of the Moon... 83
Resolving Lunar Detail with Digital Video Imaging.................... 86
Quick Project: Experiment for Effective Pixel
    Saturation with CCD Video.. 88

Chapter 8   The Lunar Terminator... 91
Quick Project: A Different Way to See the Moon 92
Quick Project(s): Finding Specific Regions of the
    Moon and Features at the Terminator...................................... 93

Section III **The Greater Solar System**

Chapter 9   A Quick Guide to the Solar System 101
The Sun 102
Rediscovering the Planets... 105
The Use of Color Filters 105
Everything Else in the Solar Realm.. 106
On Being Useful.. 107

Chapter 10   Planetary Imaging on a Time Budget 109
Drawing... 110
Quick Project: Drawing Jupiter in Pencil from Observation 111
Quick Project: Drawing Mars in Pencil from Observation 113
Quick Project: Drawing Saturn in Pencil from Observation 114
Drawing in Color... 114
Quick Project: Drawing the Planets in Color................................ 115
Some Imaging Perspectives 119
Quick Project: Combining the Best of Video and Drawing........ 119
A New Solution!... 120

Quick Project: Combining Video Frames
and Drawing - Jupiter and Mars 120
Quick Project: Combining Video Frames
and Drawing of Saturn 124

Chapter 11   Spectacles in Our Neighborhood 127
Filters, Again! 128
Quick Project: Evaluating Views of Mars, Jupiter,
or Saturn with Color Filters 128
Mars 129
Quick Project: Mapping Mars 138
Jupiter 140
Quick Project: Drawing Small Regions of Jupiter's
Disc and Cylindrical Projections 141
Saturn 146

Chapter 12   The Far In and Far Out 151
The "Far-In" Planets: Mercury and Venus 152
Quick Project: Viewing Cloud Detail on Venus 155
The "Far-Out" Planets: Uranus and Neptune 157
Quick Project: Viewing Uranus and Neptune 160
Pluto and Plutinos 160
Visitors from the Far Reaches: Comets 163
Quick Project: Viewing a Bright Comet 166
Asteroids and Minor Planets 166

Chapter 13   Daytime Astronomy 169
Observing the Sun 169
Quick Project: Indirect Solar Viewing Using Projection 171
Direct Solar Viewing 172
Quick Project: Direct Viewing 172
Observing the Sun 173
Meade ETX-90 174
Coronado PST 174
Imaging on the Run 176
Viewing the Planets During Daylight Hours 178
Quick Project: Viewing the Brighter Planets During the Day 179
Observing During Twilight and Early Morning 180
Other Daytime Prospects 180
Quick Project: Seeing Stars 180

Section IV **Deep Space**

Chapter 14   Viewing Deep Space Objects 185
Near Deep Space 188
Star Clusters 188

Open Clusters 188
Globular Clusters... 189
Quick Project: Touring Bright Clusters 191
Diffuse Nebulae 192
Quick Project: Viewing and Comparing Diffuse Nebulae.......... 194
Quick Project: Viewing Large Diffuse Nebulae............................ 195
Quick Project: Seeing Colors in Deep Space 196
Planetary Nebulae... 197
Quick Project: Viewing the Brightest Planetary Nebulae 197
Ever-Deeper Space 198
Galaxies 199
Quick Project: Viewing Detail in Galaxies.................................... 200
Novae, Supernovae, and Variable Stars.. 202

Chapter 15   Deep Space Imaging... 205
Drawing.. 206
Quick Project: Drawing Deep Space Objects 207
CCD Video Imaging with Image Intensifier 209
Quick Project: Making Images with a Digital Camera 212
A Comparison of Methods.. 212

Chapter 16   Astronomy via the Internet.. 215
The Moon... 216
The Sun and the Planets.. 218
Comets 220
The Milky Way Galaxy... 221
Variable Stars.. 221
Deep Space 221
Supernovae.. 223
Observing... 223
Miscellaneous.. 224
Robotic and Manned Spaceflight... 225

Chapter 17   A Guide for Viewing Sessions 227

Index... 255

Section I
The Busy Astronomer

CHAPTER ONE

Introduction

Life in the twenty-first century and everything that it encompasses are advancing at a rate that is truly dizzying. It seems much more noticeable in recent years than ever before. Our time and attention are under constant assault, with demands upon them that are increasing at every turn. Is such a traditionally all-consuming hobby like amateur astronomy really possible for most people these days? Most of us cannot dedicate sufficient energies and time to such an apparently demanding and intensive activity. Worse, the standard literature usually shows little awareness of this plight. Is astronomy worth pursuing if you only have an hour or two to spend, and only every so often at that?

This book will make the case that there are indeed many ways for you to participate in meaningful astronomy, despite any apparent limitations imposed by your life. Many of the strategies and suggestions given are not to be found among the more commonly ordained approaches and practices. Plus, we will discuss what equipment you *really* need, and even more importantly, what equipment you *do not* need. In having an early grasp of this, you will understand that it is better to buy what you need the first time, rather than trying to economize and then discovering that you made a mistake.

All too often suggestions as to how to get started are given to you by those who only know of one way to proceed – theirs! A little casual investigation may only make things worse, as you look at the vast array of equipment in the marketplace: a dazzling array of consumer-oriented products, all designed to grab your attention. Without the latest this or that, it would seem that you could not possibly do anything worthwhile. Perhaps you have looked through all the colorful periodicals, and there you have seen even more consumer-oriented astronomy products on display! Aside from new lines of telescopes (usually the same old designs, only with new packaging and more added electronic gizmos), there are countless new accessories

A. Cooke, *Make Time for the Stars: Fitting Astronomy into Your Busy Life,*
DOI: 10.1007/978-0-387-89341-9_1, © Springer Science+Business Media, LLC 2009

promising untold benefits (in truth, most of these accessories you can live without), new software applications (are you supposed to be more in love with your computer than with the sky itself?), and elaborate CCD imaging systems and techniques (which require a level of immersion and dedication of time that you know you cannot give).

So although we, as amateurs today, have some much-improved tools (and also some relatively new ones!) to enhance our observing potential, it is equally important to sift through the array in the marketplace and choose only those items that will truly deliver the results you are looking for. You should take great care not to allow anything to change or supplant what you really care about – in this case, those tools that will help make whatever time you have at the telescope more effective and productive.

Despite the advantages that some of the new gadgets have brought us, there is no doubt that many "observers" today have become actually more like equipment operators instead. Extended time at the eyepiece is an increasing rarity. If you value your limited time, do not be one of these! Everything else requires more devotion of time than you may have or are prepared to give.

The ranks of amateur and practical astronomers are growing quite dramatically. In the marketplace, most solutions to solve the time issue seem to begin and end at the most superficial level. A popular concept offered today is built-in and pre-programmed "sky tours" for telescopes and/or telescope controllers. The sky tours of such telescopes, which already have a "go-to" capability, involve a collection of preselected objects. They locate the objects and spend so many minutes with each before moving on to the next.

Indeed, if your curiosity only goes so far, this may keep you happy for a while. However, as you know, the glibbest activities requiring the most minimal demands of the user usually grant only minimal pleasures. It is fun at first, but soon the fascination melts away, because true insight is missing. Added to that, most preprogrammed deep space objects are so faint that they are largely out of range for the modest apertures of these popularly promoted telescopes! Thus, none of these is likely to do much to connect the users' imagination to the real wonders of space; it is more likely to send them packing. Indeed, it seems often depressingly more like flipping between television channels or playing video games, the very antithesis of good astronomy. Such capabilities do little to address our needs.

Your already overwhelmed senses may leave you with the impression that in order to get anywhere, you will need to dedicate more hours than you have left in your day, and more dollars than you may have left in your wallet. Perhaps you already know instinctively that astronomy for you, as defined within commonly accepted circles, will result in whatever equipment you have ending up in the darkness of a closet instead of under the darkness of the night sky. It may feel as if astronomy is something that will have to wait until another time, when all the cares of your daily working life have been left behind. Thus, this small attempt to show that there are indeed ways you can pursue satisfying astronomy, despite having limited time for it, or even means, at your disposal. Aside from guidance concerning best values, you will find within these pages numerous "quick projects" – activities in which you can easily take part that will bring you great satisfaction. There are perhaps many similarly time-efficient projects you might find on your own.

Before you get the wrong impression about the astronomy marketplace, it must be said that there are some truly wonderful things available that greatly facilitate taking part in astronomy on the terms we seek, making it better than it ever was before! It is just a matter of understanding what will really help us, together with knowing how to go about it. Aside from making it far more effective, these items make the hobby easier, faster, and more enjoyable. You do not have to look far to see that modern technology and manufacturing have made available larger, more consistently accurate and more affordable optics, great new eyepieces and other advanced optical designs, sizeable lightweight telescopes, electronic and nonelectronic enhanced viewing devices, the standardization and ready availability of excellent tracking capabilities, digital setting circles – to name just a few; these are all great advancements to be sure. However, in taking advantage of what truly advances our purposes, *and knowing what to leave alone,* we will find a glorious union of sorts. However, you should know that if you cannot afford to indulge in all that you desire, there are still ways to access much of it at a fraction of the cost. You can take part in great astronomy on a shoestring if you need to. Just ask John Dobson (more on this later)!

Many seasoned amateurs' astronomy "upbringing" occurred during that great era at the dawn of the Space Age, seemingly infinite with possibilities. It was certainly one of the golden ages for the imagination, even if our visions of flying cars, idyllic, futuristic, and leisurely lifestyles did not turn out quite as we had anticipated. At that time, a certain level of sophistication in amateur equipment had already evolved and was reasonably available. With some excellent products on the market the commercial supply was nevertheless still not so extensive, or so dependent upon automation and electronics, as to take all the fun away. However, commercial products were also relatively expensive. Because of the cost and limits to what was offered in the way of variety, it was normal back then for many amateurs to make their own telescopes, either from scratch or sometimes combining available components into fanciful designs.

Amateurs' instruments then ranged from the conventional (usually Newtonian reflector designs) to the most unique designs, and also to the truly bizarre. Although their efforts did not always result in top-rate or sophisticated gear, it was informative, and completely engaging in more ways than it is possible to say. It was normal to spend as much time tinkering with these "spaceships" as actually using them, but that was all part of what made it so wonderful. Amateur astronomy of the time was a curious blend of observing and telescope building, a special, fanciful place of inspiration and mystery that beckoned from the night sky.

The three volumes of the 1920's classic, *Amateur Telescope Making: Scientific American,* which detailed visions often forged into reality by many an accomplished amateur builder, figured large in all of this. If you have not perused these volumes for yourself and are yet to be fired up by these pioneers, you have missed out on a treasure trove of inspiration. If you can step back in time and see the hobby through their eyes, these volumes will change you forever. However, in this day and age it is hard to justify the kind of time required to fabricate equipment when so much is readily and inexpensively available. And most of us simply do not have that kind of time anymore.

Therefore, for the type of astronomy that is meaningful and practical for you, it boils down to just a few things:

1. The ability to take part in exciting and meaningful astronomy, with only limited time at your disposal.
2. The selection of the most appropriate equipment to reflect your circumstances, so that you will use it when you do have the chance. (After all, who would not be deterred if the process of setting things up takes too much time, offers unsatisfying results, and leaves you exhausted when you are already tired after a busy day?)
3. The ability to achieve some desirable objectives, which only large amounts of time and dedication could have brought about before.
4. Finding meaningful and realistic astronomical projects to fit your lifestyle.
5. Having an organized approach for what you do to make the most of your time.

Today, we can chase the stars in entirely different circumstances from those of earlier days, while trying to keep the old perspective alive, and by taking advantage of a far more sophisticated level of gear than we ever had, or imagined having, before. An ideal setup might consist of a modern design telescope of the largest practical readily portable configuration possible, for maximum performance for size and weight, plus powerful enhancing accessories for viewing and imaging. This will all be detailed in upcoming chapters. Such equipment would allow you to regularly experience sights at the eyepiece that sometimes equal those of many CCD images! What you will see will be live and not on a page or computer screen, instant, and not the result of hours of tinkering. It is true that it is not necessarily as vivid or brilliant as all that we have become so used to seeing from modern imaging, but the eye has unique capabilities to compensate for this. Indeed, with live observing, actual brightness and subtleties of such views will appear far greater than they really are, since the eye and brain also perform some remarkable adjustments, to a far greater degree than most people realize. The good news is that a practiced and dark-adapted eye, together with reasonably good quality equipment, *and no special accessories,* will produce results far more impressive than most images reproduced on the printed page – by any method.

It is another matter when we try to record effectively what we can see easily. The dedicated CCD enthusiast or long-exposure photographer has always had a unique turf, with the goals being to go far beyond what the eye can detect. Before advanced imaging technologies ruled the day, drawing was the amateur's primary recording method. This is still recommended as a starting point, because it teaches you to "see" in a way that no imaging method ever can. If you are not prepared to spend much time away from the eyepiece, you will want to keep your imaging simple and reasonably effective. You will probably wish such imaging methods could duplicate *only* what you see live, since you may not be interested in more complex imaging objectives per se. The methods outlined in these pages represent a growth curve through much experimentation, each method coming closer than the last, leading to ever more successful results.

One of the approaches involves using CCD video cameras. The results obtainable this way, exclusively from video "footage," are pretty good, all things considered.

When these cameras burst upon the scene, they were revolutionary; nothing like this could have been contemplated only a few years earlier. Coupled to an image intensifier, you can even image deep space subjects in real time. (Images are essentially 1/30th-of-a-second snapshots – the exposure time of a single video frame, and of faint objects in space at that!) For the most simply produced still images, the best single frames taken from the moving record require little processing. A little brightening here and there, sometimes increasing the contrast where necessary to make them look closer to the live view, that is about all there is to it.

Comments that such deep space video images do not always compare to the enthusiastic descriptions accorded to them have often been made by those who seem unaware that instant video images of deep space destinations were not feasible previously, by any method! However, it is true that this form of deep space imagery, while producing remarkable results in real time, nevertheless has significant limitations. These are all too apparent when reproduced on the page.

Searching for better ways to proceed, you might look at frame-integrating CCD video. Images that you may have seen produced by such cameras certainly offer far improved results, along with at least some degree of simplicity. However, this process still requires a lot more trouble and hassle than you may be prepared to give. There are still real limits in showing of some of these faint and delicate subjects because of the finite lines of resolution imposed by the video system itself. Subtle though these may seem, they ultimately detract from the feeling of the live view.

Ultimately, various paths address most demands for easy imaging – lunar, solar system, as well as deep space. You may be quite surprised by what actually is possible using the simplest and quickest approaches. What is reproduced on the page appears actually much closer to the subject's appearance in the eyepiece; you can judge these for yourself. The best part is that you do not have to become a techno-geek!

The following chapters will examine more closely the specifics of these imaging solutions. Hopefully they will work for you; although they usually still do not equal the best images produced these days by advanced imaging methods, they do represent a giant leap forward and succeed in providing remarkably good visual likenesses of the space objects themselves, especially as they appear in the eyepiece in general conventional viewing. Suffice it to say, the main purpose, therefore, for their inclusion in this book is to serve as a general guide for what you will see through the eyepiece in moderate and larger apertures under favorable conditions.

CHAPTER TWO

Making the Right Equipment Choices

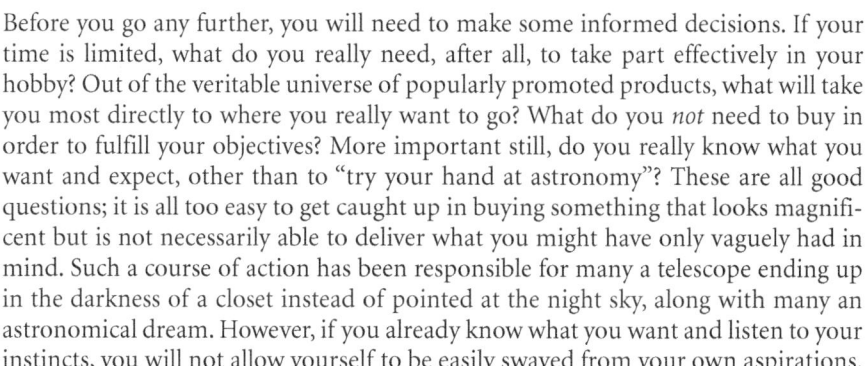

Before you go any further, you will need to make some informed decisions. If your time is limited, what do you really need, after all, to take part effectively in your hobby? Out of the veritable universe of popularly promoted products, what will take you most directly to where you really want to go? What do you *not* need to buy in order to fulfill your objectives? More important still, do you really know what you want and expect, other than to "try your hand at astronomy"? These are all good questions; it is all too easy to get caught up in buying something that looks magnificent but is not necessarily able to deliver what you might have only vaguely had in mind. Such a course of action has been responsible for many a telescope ending up in the darkness of a closet instead of pointed at the night sky, along with many an astronomical dream. However, if you already know what you want and listen to your instincts, you will not allow yourself to be easily swayed from your own aspirations.

Certain fundamentals would seem to be constant. Once bitten by the astronomy bug, who among us has not been entranced by the alluring appearance of an astronomical telescope? A telescope seems to exude the highest expression of scientific adventure! How many of us have spent time early on imagining all of the adventures in the universe we might experience via a backyard telescope? Such dreams always conjure up all kinds of fanciful rigs, and the telescope itself may have often figured nearly as large as the activity it was intended for. There is nothing wrong with enjoying these astronomical tools just for themselves, for all that they represent, beyond being merely a means to an end, including the wealth of astronomical history and personalities associated with them. Their mystique always seems intrinsically wrapped up in all of this.

What is ideal for one person is not necessarily so for another, and therefore the right choice for you requires just a little thought. Perhaps you are just starting out. In this case, either the sky or your bank account is the limit, and there are many

A. Cooke, *Make Time for the Stars: Fitting Astronomy into Your Busy Life,*
DOI: 10.1007/978-0-387-89341-9_2, © Springer Science+Business Media, LLC 2009

potential options to consider. Beware of any telescope on display in a store at a shopping mall; it may look impressive, but it is more likely to be of little use beyond the curiosity of the beginner's first ramblings among the stars. And if you are already an old hand in astronomy, it is not necessarily a prerequisite to reject or radically upgrade the equipment that you may have had for years. The newest technology will not reduce the effectiveness of anything that already works, but you may still be able to get more from your equipment– even a lot more – to make better use of your time. In other words, some of the newer equipment can enhance and streamline the type of astronomy you may have enjoyed over the years while not radically changing it. What about affordability? You can still access a large part of the whole with significantly less than the ultimate in equipment described here, so financial constraints need not necessarily restrict your dreams. However, there is no way to protect you from craving ever-greater telescope apertures and better accessories. This is the astronomers' incurable disease!

If the subject of this book lured you to it in the first place, it is possible that you may be aspiring to take part in something utterly different to all that we see so widely propagated today. You may not consciously know it, but you may already have an aversion to so much of what we see as consumer-tech dominated astronomy, with all of its corresponding auxiliary equipment (implying heavy time demands), and all seemingly promoted as necessities. Do not think that casually finding something that will resonate with you is something you can take for granted; today's mainstream agenda may not coincide with yours, or the attention and proportion of your life you are able to give. You need to put aside the pressure from anyone else's predetermined vision of amateur astronomy, something visible in much of the amateur astronomical media, and which seems to march lockstep with all of the latest commercial developments. Thus, instead of unwittingly accepting a substitute for your aspirations or what they might have been, hopefully you will find something more in line in them. Certainly there is still no shortage of the right astronomical gear to help you make your own custom fit.

From Absolute Zero to Absolute Minimum

When it comes to equipment minimums, some old purists would argue, and not without some justification, that all that we really need are our eyes, mind, and knowledge of the sky. After all, the great astronomers of antiquity had to manage with barely more. Other devotees of the simpler approach would argue that just having a good pair of binoculars would complete their needs; indeed, many amateurs have had a lifetime of enjoyment with little more.

Although it is hard to fault the pure quality of such thinking, for most of us this simply will not be enough. Just knowing there are ready means available to transform our experiences as observers will be enough to push most of us in a more equipment-oriented direction.

Potential Pitfalls – Do Not Fall In

If you are trying to avoid buying useless equipment, or spending large amounts of money on features you will never use or soon outgrow, there are lots of potential pitfalls. Frequently fancy features are supplied at the expense of effective design, so beware. Not long ago a magazine article, when referring to conventionally mounted instruments, actually used the term "push-to" telescopes! The writer was presumably straight faced. Here, we have the new way to describe anything that does not comply with the much ballyhooed, but hardly necessary, "go-to" telescope, apparently the new sanctioned standard. If you were to adopt this line of thinking you would conclude that anything less means something inferior, or worse still, obsolete, and no longer of value. (The *go-to* concept was originally designed for massive observatory telescopes, whose huge and cumbersome dimensions required considerable skills, to say nothing of the strength and patience of the operator. Perhaps some truly monstrous amateur configurations would qualify as good candidates for such automation, but not the diminutive little scopes we often see on the salesroom floor.) In the marketplace of amateur equipment – aside from some notable exceptions – *there is absolutely no need for "go-to" capability.* Do you want novelty or telescopic performance for your money? And if, perhaps, you should want novelty as well, is it worth as much to you as possibly the scope itself? Save your money. The exceptions to this, however, might include some of the shortest and stubbiest of all optical configurations (and least valuable to visual observers). These designs, it is true, are indeed sometimes more awkward than some less compact configurations to position easily and accurately by manual guidance alone.

You may find it hard to avoid the perception that we must invest in quantities of gear, which often includes needless electronics, and *especially* the dominance of imaging capabilities. All too often these telescopes come part and parcel, sadly, with the somewhat compromised optical designs that dominate the marketplace today. If the powerful force of commerce has left you feeling that your own needs are quite different from these, you may already be developing insight and your own astronomical vision.

When starting from scratch it is important to recognize the virtues of having the maximum aperture possible. Forget about magnifications offered; these are exclusively a factor of aperture and nothing more. Just because the box says that the telescope is a 750× instrument it does not mean that the image seen through it at such a power would be anything you would want to see! The telescope should be portable, should it need to be transported, and, it should be of high quality. And always remember, maximum possible aperture is the key to everything you do. If you are a suburban dweller, you will want to observe primarily from your home location, since time is a factor. Do not give any credence to that old humbug decreeing that larger apertures are of less use than smaller ones from these locations, or even of limited use in such environments. Only their full potential is limited, but in decent conditions they win every time against lesser sizes, regardless of location! Good optical science has shown this conclusively, as anyone familiar with the practical application of it already knows.

Although aperture and the advantages it brings are of universal value (the only downside is their slower cool down periods), it is also true that the larger sizes only really come into their own under dark skies, and quite disproportionately so. Their potential performance in such favorable conditions is far greater than that of lesser sizes, and it is in these surroundings that their full capabilities may be realized instead of just glimpsed. Thus, always keep an eye to taking your astronomy to great locations, even if your opportunities to do so are limited; telescope size is thus quite significant when it comes to choosing the right telescope.

It is also possible that you may indeed elect to buy something that coincides with today's most commercially promoted parameters, for various reasons of your own. There can be no quarrel with that. It is up to you to decide what suits you best, after all, but do it as an informed consumer, and try to buy only what you truly need for your own purposes. Indeed, most complex features will not be even particularly valuable for the majority of people, unless, for example, CCD imaging by remote control, or some such elaborate option, is going to be your thing. Meanwhile, many features may be more akin to those on many common modern appliances; they look good on the device itself and make for a great sales promotion, but few of them will ever be used. If you have arrived at the conclusion that any of the latest trends in amateur astronomy, such as CCD astronomy, is what fires you up, there is no reason that you shouldn't follow this direction. However, be prepared to spend more time with your hobby than you might have bargained for, and be sure that whatever you select is of your own choice.

If you already have good equipment and wish to hang on to it, you can still take advantage of many of the more useful accessories in the marketplace, equipment that enhances what you do and ties in more readily to the kind of astronomy you want to take part in. This is the best of both worlds. Many of these accessories and devices will fit right onto your original instrument, or work beautifully in conjunction with it.

You should also try to avoid settling for something that limits your potential, and which will only lead you back to the marketplace again before long. Remember that aperture and quality are both important here; there are plenty of less than excellent telescopes in the marketplace of respectable apertures but which are almost useless for any sustained application to astronomy. If you want to have equipment that you will not soon outgrow, then for solar system viewing the smallest aperture you should consider would be around 4 in. (10 cm) for a high-quality refractor, or 6 in. (15 cm) for a good reflector. However, given a choice, the reflector wins, because of its greater light grasp (hence potential with deep space subjects), greater ease of use, clean imagery, and price advantage. Surprisingly much, if not most, physical detail likely to be visible on solar system objects will be apparent with sizes only somewhat bigger than these minimums. Although greater sizes do indeed add to the ease of viewing and increased resolution of detail, most of their advantages beyond those of somewhat smaller sizes will be in the discernment and prominence of colors, as well as other even vaguer subtleties.

For deep space, it is a different matter again, and ideally you probably should not consider anything less than 6 in. (15 cm) for a refractor and 8 in. (20 cm) for a reflector, although a good 6-in. reflector, or even a 4-in. (10 cm) refractor, is far from useless. These sizes used to be the amateurs' ultimate workhorses. Larger sizes yet are noticeably better, since for viewing faint objects, scooping quantities of light is the name of the game.

The reflector would seem to be the king for all of these objects, as few amateurs will be able to afford really large refractors of quality. Besides, ever-larger primary lenses bring with them a host of other problems.

Although the humblest to the most grandiose commercial optics can attain surprising quality these days, the most popular compact and portable optical configurations (catadioptrics) dominating most manufacturers' catalogs do have significant downsides in performance, regardless of quality. Being neither pure reflectors nor refractors, they need to be somewhat larger in either viewing category, and preferably substantially so, since they are the least "light efficient" of the bunch. As a ratio, read at least 8 in. (20 cm) of good quality aperture for a catadioptric to match 5 in. (12 cm) of refractor aperture in all categories of viewing. In some cases, this is overly generous, and even then, the contrast they offer is still likely to be inferior to other types of telescopes. Despite their proliferation in the marketplace, these are the simple facts of the matter. This is particularly the case in regard to the live view they provide, which suffer, by default, from some degradation due to these telescopes' inherent optical design.

As if to draw attention away from this less than desirable attribute, most of these commercial telescopes seem to tout (actually, they "scream") electronic sophistication and gadgets over actual viewing! Just look at any advertisement; if you did not know better, you would wonder how any of us got along without all the features their instruments seem to boast of as being key items. Additionally, CCD imaging has resulted in an emphasis on something other than the pleasure of simply looking through a telescope. Certainly CCD (and the complex processing that comes with it) overcomes most of the ill effects of very compact designs, but it presumes that such an indirect use of the telescope, as opposed to live viewing, is for everyone.

It is entirely possible, of course, that your own circumstances will dictate that you ultimately consider something of a compromise in optical configurations. You may like the compactness of a catadioptric telescope, even some of its electronics. Ultimately, practicality may dictate something that adequately fits the bill overall, and while, in an ideal world, what you choose might not have been your first choice, it will still allow you to pursue things essentially in your way.

Nevertheless, regardless of choice, try to steer away from needless technical complexity, especially when you could use your hard-earned funds instead for better quality, greater aperture, or really useful accessories.

What about telescope types and value for money? The venerable Newtonian still offers by far the best value for the money, and its ease of use should keep it high on any list. What about all of the supposed hours this design of telescope requires for maintenance? Humbug! Today's Newtonians have largely made these criticisms irrelevant, at least when using a reasonable-sized aperture. (Be careful of smaller examples whose optics and flimsy build are not likely to live up to their promise.) The smaller sizes of Newtonian are more likely to fulfill their role best if their focal ratios are F8–F10.

However, for comfort of use alone, it is more likely that the larger (and shorter focal ratio) Dobsonians and Split Ring Equatorial models will be preferable, especially if the observer is often able to view from more natural standing positions and with minimal reach. Because a practical and comfortable eyepiece position is highly desirable with any telescope, just be sure that any Newtonian you are considering stands high enough, and that with an equatorially mounted instrument, it offers a simple way of rotating the tube or eyepiece to maintain a reasonable viewing position at all times. Otherwise, you will wind up in situations where the eyepiece is on the underside or some other awkward place and find yourself stooping to look in the focuser – just what we are trying to avoid! An eyepiece that is placed, by default, at a comfortable height and viewing position requires no awkward bending or stretching, and there is no need to look up and underneath the telescope, as with the refractor. This is the optical configuration most beginners automatically associate with that of a "telescope." Physical strain seriously detracts from the pleasures and efficiency of observing. Similarly, with a Newtonian, there is no need for a star diagonal in order to overcome the shortcomings (along with its reversal of the image), and with the larger sizes, there is no need for inordinately high mountings or wide-footed tripods (both with the potential for instability or tripping on them in the dark).

However, there comes a point where ever-increasing size does begin to present its own problems, necessitating high viewing platforms or unstable ladders just to reach the eyepiece. For most of us, though, there is a happy place somewhere in the middle. Most observers would probably say that an ideal-size Newtonian would range between 12 and 20 in. (30–50 cm), with shorter focal ratios dominating the larger sizes. Be sure to read "Getting more from Your Newtonian" in Chap. 4 before making any final choice.

Now let us talk about apochromatic refractors. All of the euphoria and praise you may have heard about them is appropriate, at least as far as their optical performance is concerned! However, the awkwardness of their use when viewing overhead objects, coupled with their prohibitive price relative to aperture, actually makes them one of the worst choices overall! So although it is true that they do give maximum image quality inch for inch, the drawbacks associated with them mean that they do not necessarily provide the best value for your observing dollar. Most observers will never be able to afford a truly large apochromatic refractor, since, because of the high relative cost, the size you choose will be much smaller than what would be considered a large reflector. Indeed, most examples we see being used today average 4-in. aperture (10 cm) or less. Therefore, it is far preferable to invest in something that will give you instead much more viewing for your money, with nearly as much optical perfection – perhaps not quite as aesthetically pleasing as the solid touch of a high-end precision refractor or quite so close to reaching its optical perfection but an instrument that will actually serve you far better in the long run. Buy something that can deliver large amounts of well-focused, minimally scattered light comfortably and stably to your eye through good mechanical and optical design. Such qualities are the most important ingredients in that special formula for enjoying productive, time-effective amateur astronomy. It usually comes down to Newtonian designs, in many ways the simplest. Many people do not realize that the most straightforward optical and mechanical configurations usually prove themselves best in this regard. Indeed, half of your viewing pleasure will come from ease of use. When your time is short, this is more important than it may seem.

Among the various Newtonian designs, the now standard and well-accepted Dobsonian telescope (Newtonian in optical configuration), is the king of the value kingdom especially since, in its fundamental and basic form, it is about the simplest and cheapest concept to buy or build. Just a standard Newtonian set on a massively large altazimuth mounting, it nevertheless takes advantage of both axes being at the lowest possible point, with the weight of the primary mirror within both of these axes and few or no counterweights being needed. This provides stability and great ease of movement, and the axes may be allowed to be quite stiff, something indeed preferable. No need for beautifully engineered frictionless bearings here! Tracking objects by hand is not especially difficult with Dobsonians, which are also very quick and easy to set up. Even if fitted with digital circles or a tracking platform these remain among the best choices for someone with a serious shortage of leisure time. Most serious devotees of the design are likely to be live observers, so do not look at it as being any type of compromise. The whole idea was to find a way to provide substantial aperture, stability, and low cost in one package, something that came out of unique circumstances. This breed of scope (the name of which now pays tribute to its innovator, a former monk named John Dobson) was the result of his having no other options during the time that he began telescope building. For him it was either doing it the only way available at that time, or doing no viewing at all! Thus, he found his solution out of necessity, with no means to take any other route. He must also have had very limited time, doing most of his building during the late hours when no one was around! Some of the descriptions of the components he used to build his optics and mountings would make an engineer blush, especially anyone with a background in telescopes! (Actually, they would make *anyone* blush!) However, it was never truer that one man's junk is another man's treasure, because principle clearly triumphed over aesthetics.

Dobson's concept soon became dominant in the large amateur aperture league and has become a familiar sight at any observers' star party. The San Francisco Sidewalk Astronomers (the original group with whom he will always be linked) made these telescopes legendary and featured what was then the largest amateur telescope in the world, the 24-in. "Delphinium." If you consider what else was available in the amateur world back then (the late 1970s), you will realize how remarkable this was at that time. Unwieldy and anything but easy to use, it ushered in a new era of giant amateur apertures and demonstrated that such large instruments were not reserved just for professionals. Dobson had realized with supreme clarity at that time ultimately that what we can actually see (and see easily in fact) is still the core of many amateurs' astronomy today! Everything else was well down on his list of priorities. Thus, discovering what can be done with the simplest and least expensive approach is particularly encouraging in an age that urges us to spend seemingly limitless amounts of money on ever-fancier hi-tech gear and devote ever-increasing amounts of time to applications, after the fact, from which we may already feel disconnected in the first place! Our astronomical roots, as exemplified by the original amateur telescope makers of Springfield, still mean that we can indeed have something worthwhile from unlikely and simple means! But look closely; you will see that Dobson incorporated the most important features from Russell Porter's timeless designs of the 1920s. It seems that one or two people knew these things all along.

By incorporating high-grade components to make more sophisticated instruments of this type, Dobsonians may rate very high in satisfying many fussier observers' requirements, while still being the easiest form of telescope for the amateur to use, set up, or build. Thus, it may indeed be realistic to undertake building one yourself no matter how ill equipped or ill suited you are to handle mechanical things. And these telescopes remain decidedly cost effective, even if you *buy* all of the optics and fittings.

However, because of the basic Dobsonian's emphasis on simplicity, coupled with stability and economy, the optics in them may not always be of the highest order, though typically even the lesser examples will be found to be acceptable for most general viewing purposes. However, low-end instruments may be more suited to wide field deep space views, where light grasp is the name of the game, than in revealing the many subtleties of planetary detail. This need not be the case if you are prepared to pay for a better, or even top-of-the-line, Dobsonian with the best optics and mechanics. Obviously, the solid-tube versions are not likely to be as portable as truss-mounted varieties, but at least their low-slung mountings do not preclude moving them around with a reasonable degree of ease. They are usually supplied with carrying handles. Some of them, including the very best of the commercially built varieties, have made available some awesome apertures, the likes of which used to grace only professional observatories. Coupled with tracking capabilities, the uses of these top-of-the-line rigs are practically unlimited.

Relative simplicity, time-effective easy setup and use, and maximum possible performance, with the most direct and immediate visual results – what is not to like? By taking advantage of technical advancements to enhance this approach, you will be able to participate in some satisfying visual astronomy even from suburban locales, together with being able to readily transport this equipment to other locations for positively spectacular viewing!

The Homebuilt Telescope

Dobsonians certainly can be put in this category. However, many unique "one-off" instruments can be built by enthusiasts, ranging from the most sophisticated to the truly primitive. Some will impress even the most jaded or disinterested party. But primitive is not necessarily a bad thing, either. It is a curious fact that some of the most memorable times can be had with the most basic and limited equipment. The lack of having anything sophisticated may even impart a greater sense of adventure to a developing interest than having immediate access to the typically automated products of today. The fun may only be enhanced by building something yourself, or modifying another scope already in existence.

A good example of this author's own homebuilt aspirations included what seemed like an astounding aperture at that time: a 12½ in. (31 cm) F9 Newtonian reflector of 1977 (Fig 2.1). It was designed specifically for planetary viewing and had F9 optics with the resulting long and nearly unwieldy tube length (!), a tiny secondary mirror more like a secondary for an 8-in. (20 cm) in order to provide the best contrast possible, together with a horizontally sliding focuser and secondary mirror mount to move the eyepiece and secondary laterally along the length of the tube, which, in turn, kept the eyepiece always as close to the secondary as possible. Built completely from scratch, it functioned beautifully doing what it was designed to do, although it did take a lot of time to build. However, even at that time, amateurs had many other options widely available in the marketplace, so you would conclude correctly that in those days the joy of building, as well as real affordability ($300 at that time), was an important part of the whole as well as the use of the telescope itself.

The unique design paid large dividends. For planetary viewing, inch for inch, views through it were more like a fine refractor, except without image color fringes. Never mind that it had no motor drive, electric focuser, setting circles, etc.! This only reinforces the point about what you *really* need in order to have great adventures in astronomy. The perspective gained from such close hands-on building experiences seems lost forever to all but today's most hardheaded traditional enthusiasts. With the abundance of affordable and sophisticated instruments on the market today, it is much harder for the present generation of amateurs to drum up enthusiasm for building something these days, considering the challenge and the likelihood of inferior performance. Nevertheless, special optical configurations and an unconventional mindset for experimental designs can occasionally still provide justification enough to return to the do-it-yourself philosophy for some. However, if this is to be your thing, you will need time to spare. Again, that may rule it out for you.

When you have made virtually every component yourself, there is something close to complete disbelief upon initially peering into the eyepiece and seeing a distant landscape dazzlingly realized. You can hardly believe that the confounded thing, your creation, actually works! Although most readers of this book will not take up building their own, perhaps borrowing just a little from the mindset of the amateur astronomer/telescope builder/tinkerer will provide some insights into adventures unknown to many enthusiasts today. However, the main point in all of this is that you can have a very good time indeed with quite minimalist, if not exactly minimal, equipment.

Sadly, the better examples of larger telescopes from our collective not-so-distant yesteryears, either of the homebuilt or commercial varieties, will probably have limited practical value to most users today. It is not that many survivors are lacking in quality, or are insufficiently advanced in design to give first-rate service; they are often among the finest ever made. It is simply because they are big and heavy! It is a depressing truth that many of these wonderful but relatively massive and bulky old designs may not be moved around easily. Portability has become a prime ingredient in the mix. An observer who actually lives under dark skies could put to use almost any good quality telescope design, type, or age. Fortunately for these people, some of the best quality scopes and designs ever commercially built date from not-so-distant earlier times; it may be possible to find one for a remarkably affordable price. (None of the "new astronomers" have any use for them!) Many of these grand old scopes happen to be among the most aesthetically pleasing of all, as well. At the extreme end of this scale, a few years ago one of the fabulous 12-in. (30 cm) telescopes designed by Russell Porter for use in selecting the site for the Mount Palomar 200-in. (5.3 m) telescope came up for sale privately. Although truly massive and surely far from easily portable, for the lucky buyer it would have represented one of the finest 12-in. telescopes ever built. Lucky indeed was the purchaser!

If you happen to live "out in the boondocks," building something yourself is a much more realistic option than for those who live in or near a city. It is much easier to produce something good that is rather lengthy and massive than it is to build a more compact instrument of comparable performing value. Thus, it may be more realistic if you never have to move your scope to a better observing site. Longer focal ratios are unquestionably far easier to build than anything of a shorter focal ratio. So, if you are fortunate enough to live somewhere with superior viewing conditions,

Fig. 2.1. The author's 12½-in. (31 cm) F9 homebuilt telescope of 1977. The large bearing surfaces of the equatorial mounting were built from large plumbing "T" fittings, hand machined on the insides with an electric drill and grinding wheels, enclosing rotating concrete-filled, heavily greased 6-in. iron pipe axes! These provided exceptional rotational smoothness. Very little clamping of each axis was needed to hold the tube exactly where it belonged. Tracking movement was easy to control by hand, though a motor drive would have been better to eliminate hand-inducing tremor and to keep things always centered.

and no need to transport your scope to somewhere else, you would be well advised to carefully study and heed Russell Porter's design principles in *Amateur Telescope Making*, Volume 1. Nothing would give finer results than realizing any one of his designs, which he so ably and imaginatively illustrated.

Why is it that we cannot readily just set about building more compact designs for ourselves? To begin with, making really good larger optics, especially those of short focal ratios, will tax even the most skilled amateur optician, probably to a greater degree than most can overcome. Compounding matters considerably, the larger the primary, the greater the difficulty in perfecting the surface; the physical task alone of grinding, polishing, and figuring the optics presents an increasing challenge as the aperture increases. The problems compound in a steeply increasing gradient. Most amateur opticians simply lack the necessary skills, equipment, patience, time, or even the necessary elbow grease to produce quality large optics, let alone the smaller primaries of more limited apertures. Grinding and polishing machines can indeed help matters considerably, and some of the most ingenious telescope builders of the past have produced their own. However, designing and constructing such machines will again push the challenge curve into the realm of impossibility for most of us. And it will take more time and ingenuity than most people have at their disposal. We have thus gained something, but not without the loss of something else – in this case, the wonderful experiences of the amateur's traditional dual role of astronomer–builder. So, while scratch building portable short focal length telescopes is not impossible, most enthusiasts will probably elect to buy something that is portable and can be transported to a good site once in a while, even if the design of such a telescope would not be their first choice.

In addition, building mechanically sound portable assemblies from sufficiently lightweight materials will also create stability challenges greater than many amateurs will be able to address. Stability is the Achilles' heel of many amateur designs, and it is easy to recall a fair number of such telescopes, which looked massive and wonderful to be sure but had all the sure-footedness of a drunken sailor. Of course, all of these problems can be overcome with enough skill and knowledge, as any attendee of a large star party can attest. Some remarkable examples of masterful amateur construction do indeed exist.

One notable, if not notorious (!), example of a profoundly nonportable mounting is shown in *Amateur Telescope Making*, humorously, and almost reverently, nicknamed "Porter's Folly." In this design, seemingly ridiculous amounts of concrete and sheer mass are employed to produce perhaps the most stable mounting ever created for an amateur scope. If ever built, it would be a marvel to experience in practice! Porter *almost* overemphasized such stability in his descriptions; it was not enough that a mounting was rigid in the normally accepted terms of engineering. It had to be *inflexible*. This included the requirement that any vibration was to be damped almost instantaneously. Porter understood and advocated for such stability probably better than anyone has, before or since. Although his designs may not always be pretty (although they certainly are always of handsome proportions), they do offer the builder a practical value and understanding of mechanics few of us will ever fully know. Strikingly, these designs and visions emanated long before World War II. Those familiar with them already know that their unique value has not been diminished with time.

All of this underscores a sacred principle you should not overlook when making your telescope selection. Although you probably need to look for portability, you will find that many telescopes built today incorporate Porter's design concepts, but now with an eye toward lightweight construction. Among them is the so-called *Split Ring* mounting design. The range of JMI telescopes (Denver, Colorado) may have been the first commercial application of Porter's design outside professional observatories. And the mounting selected for the 200-in. Hale Telescope on Mount Palomar just happens to incorporate most of Porter's design, even if it is not strictly a "split-ring" design. Unfortunately, many instruments are still made "in a vacuum," as if Porter had never existed. What more can one say?

Eyepieces

Easily overlooked at the initial stages is the necessity of having some decent eyepieces to provide a sufficient range of powers. Every object in space has its own optimal power, which is further dependent on atmospheric conditions. You will also need eyepieces sufficiently comfortable to use over long periods of time. Great enjoyment may still be had from eyepieces that would be regarded as archaic today; they are actually a lot better than one might imagine. It is quite possible and realistic to do well, even now, with such an assemblage of older eyepieces, and if you can find them used, which might include some Kellners, an Orthoscopic or two, and say, one good wide field Erfle, you will save a relative treasure chest of money to boot. The last two eyepieces mentioned, while not quite as advanced as the best modern designs, will still deliver impressive performances, even by direct comparison. It is fair to say that you should try doing better than to rely on anything as primitive as a Huygenian or Ramsden. These antique designs have long had their day, with their small fields of view, limited eye relief, and achromatic problems.

The cost of the best modern eyepieces may be high, but at least we have some truly magnificent options available today that long ago would have blown away our minds. Massive and elaborate, typically very expensive designs, nevertheless they are well worth the cost if you can afford it. If not, many simpler modern types, as well as fine older designs, are still being made today and can do remarkable service for considerably less money. More on all of this in the next chapter.

If your budget allows you to reach a little further than the absolute minimum (given the acquisition of at least a decent, if not an expensive, telescope), then maybe having more than a basic eyepiece collection would make a lot of sense. A small collection of eyepieces, combined with some similarly high-quality Barlow lenses, can yield results equal to many eyepieces. With, say, just three eyepieces, you can see a lot with just a standard Barlow 2× lens. Such a lens will double the power of each of your eyepieces, but take care that your collection features eyepieces that do not have the same increments of increased power as does your chosen Barlow! Having *two* Barlows of very different focal lengths, one high power and one standard power, would give tremendous capability, especially if the Barlows are "stackable." If high quality is on your side with every optical component, you might even try such "stacking" of two Barlows together for ridiculous magnifications on those all too rare near perfect nights, but only with suitable subjects, of course! Suppose you are straining to make out detail on, say, one of Jupiter's moons on one such extraordinary night; you could even try for 1,000× while taking advantage of a relatively low power eyepiece to provide a wide apparent field. Thus, in this case, the use of a 2× Barlow and, say, a TeleVue 5X Powermate would give a *tenfold* boost. With such doubling of Barlows any given (1¼-in.) eyepiece can become effectively three or more eyepieces and certainly represents the best value, if you are looking to limit your investment. You will also get improved eye relief from the use of a Barlow in the optical train, although each added lens component does soak up some of the transmitted light.

A 2-in. (51 mm) wide field eyepiece requires a giant 2-in. (2×) Barlow (those by TeleVue are known for their superb quality) to double its power without causing vignetting (the undesirable "clipping" of the field of view). The giant Barlows usually come with a 1¼-in. (31.75 mm) adapter for smaller eyepieces as well. With the use of ever-shorter focal length eyepieces, the dimensions revert to the standard 1¼-in. size, so you can then easily switch to higher power when opportunities present themselves.

Because the quality and design of Barlow lenses has improved so steadily over the years, you should probably avoid all but the newer ones, and whatever you select should be of the highest quality available, so as not to detract from the performance of the eyepiece itself; lesser examples may leave you disappointed all too often. Modern Barlows work just fine with older eyepieces, too. Of course, 2-in. eyepieces are less flexible since they may only be used in conjunction with larger 2-in. Barlows.

Because advanced and sophisticated eyepieces are so numerous today, we will save further comment on them until the next chapter. Exactly where one observer draws the line as a minimum, what one considers a necessity and what another feels is a luxury is hard to quantify.

Right on the Money

Now for the money talk. How much do you really need to spend? Surprisingly, it turns out not to be a fortune, although it has to be realized that ultimately there are very few *truly* inexpensive routes to take in amateur astronomy, unless you make everything yourself. So, by now it will have become obvious that to follow even the simplest form of astronomy requires that you spend *something*. And it is not much use settling for anything woefully inadequate. However, there are still some viable options. Even if you proceed to buy *everything* you need, it is possible to have a lifetime of enjoyment with nothing more than the basic items, and an investment unlikely needing to be greater than $1,000–2,000. For this, you can probably find something in the 10-in. range or larger, such as a new Dobsonian, or something else, maybe even a fine used equatorial classic. Meade has been marketing a nice series of truss-tube Dobsonians ("Lightbridge" series), which fall comfortably into this category. The truss design helps greatly in portability. Orion (of Santa Cruz, California) also has earned a good name for inexpensive Dobsonians. With smaller sizes and the simplest types of all, a further amount may yet be shaved off this figure. Many of these lower cost telescopes will be unlikely to be motor driven (true "push-to telescopes").

Hopefully, you will find that it is possible to take part in real astronomy, even if you do not have much time and are on a budget. Naturally, the amount of money you spend can run many multiples higher, depending on the quality, degree of mechanical sophistication, size of aperture, and range and type of accessories you acquire, but at least there are a host of realistic options for almost every pocketbook. If your own budget permits, and you want nothing but the best, it is quite easy to rack up tens of thousands of dollars (or more!). At the lowest end of the scale, homebuilt telescopes can produce the greatest savings of all, of course, depending on design and construction as well as to what degree of its construction was built from scratch. It is also possible with homebuilt telescopes to go to the other extreme, of course, since some otherwise unattainable and magnificent instruments have been the result of enterprising and skillful amateurs; examples exist that are at a professional level, and at professional level costs, too. Regardless, all homebuilt designs come with real satisfaction as part of the bargain!

Aperture Fever

After a few or many "warm-up" telescopes, you may find yourself, like so many others, falling prey to "aperture fever." This incurable disease is highly contagious and will wreak havoc with the hardiest souls, so consider yourself forewarned! You may find yourself obsessed about the "ultimate" setup, all of which may result in an instrument and sophisticated accessories that may only be limited by space to store them and finances. Ultimately, these conditions will determine where you will draw the line, and at least this form of reality check keeps aperture fever under control. However, until we reach this point, nothing quite answers the cravings better than actually acquiring greater apertures or grander designs, since the gains for the observer grow almost disproportionately as the specifications expand.

There are various commercial options available that, though not exactly curing the fever, get it under control. Seemingly bulky instruments, such as the equatorially mounted JMI Newtonian 18-in. (45 cm) telescope of F4.5 ratio, are, in fact, the most practical choices overall, contrary to all the impressions you may have of larger and theoretically unwieldy telescopes. The design of this telescope addresses most portability needs, especially for an instrument of its size and type. It is sufficiently stable (though certainly not so rigid as a permanently mounted, much bulkier telescope of yesteryear) for most uses, and its total weight is only about 250 lbs (550 kg). So, fortunately, some of the better and more specialized commercial manufacturers, along with newer construction methods and materials, have made the unthinkable of the past a reality today. Similarly, a number of innovative independent companies are pushing the envelope toward answering the portability challenge without sacrificing optical design. Not quite able to deliver all of the contrast of a bulkier telescope with a larger focal ratio (because of the size of F4.5's secondary mirror), the JMI scope performance remains excellent and right on the cusp of where significant image degradation would begin. This, in spite of the secondary being a full 22% of the diameter of the primary; it is still small enough that one would be hard pressed to tell the difference between it and a telescope utilizing a secondary of somewhat smaller dimensions.

Other Worthy Accessories

Setting circles makes such a difference that they have become almost necessities these days, especially if you are observing in light polluted skies. You will find them such a convenience and pleasure to use that you would be wise to include them in your telescope choice, or add them as soon as possible! (Those traditional, engraved circle types, "gracing" many a small aperture refractor, are useless because of their small size, but at least they look good!)

The most useful circles to us, by far, are those of the digital variety. Their cost, though not exactly minimal, is still far from excessive, and a maximum of several hundred dollars should be enough for most types in the marketplace. Do not be swayed by those "sky purists" and snobs in the amateur astronomical community who do not believe anyone qualifies to practice the hobby without detailed knowledge and familiarity with virtually every star in the sky! How quickly these individuals regularly condemn such wonderful aids and effectively try to keep their domain all to themselves. Given today's overly automated approach, which has been embraced by so many people, it is unclear just how one is expected to gain the kind of familiarity "required" by the purists of the night sky. We also should not forget that most modern additions to the marketplace boast the very same automated features, which by default make the acquisition of such advanced sighting skills highly unlikely in the first place! And let us also bear in mind that elitist attitudes sometimes come from the same folks who seem to cheer the advent of every new commercial telescope, equipped with virtually every aid for eliminating the need for such sighting abilities! This would seem to be the ultimate contradiction. Although such a familiarity in navigating the sky is indeed desirable, it is far from a prerequisite to enlightened viewing. Just arm yourself with a general familiarity of the sky, so that you are not left bewildered at the vault above you. This should include the major constellations and most of the brightest stars.

You will also probably need some form of sighting finder scope, although you may merely line up targets in the sky along one of the telescope's tube. Although not the easiest thing to do, this crude system works quite well once you develop an appropriate technique. However, you might not even think of utilizing such a primitive method! Although a standard refractor-type finder is fine, say, of about a 50-mm (2-in.) objective with an illuminated reticule for easy star alignment, there are other good options now available. An inexpensive new type of finder projects a red dot against the sky as you align it with the object being located. These outstanding hi-tech, zero magnification devices make object location and centering extremely easy, more so than it sounds here. Indeed, they are every bit as good as anything you might use, as long as light grasp for locating faint objects is still not one of your requirements. If you elect to go the red dot finder route, buy the most robust version available, such as those marketed by TeleVue, the reason being that the slightest jolt is sufficient to throw the cheaper plastic varieties out of alignment.

These days, digital circles have, for the most part, made optical finders nearly obsolete, the primary use of a finder now being the locating and centering of alignment stars at the outset of an observing session. Once this has been done, the only other use for one is in centering a solar system object or some other bright subject;

certainly, there is no need for large finders or setting circles to sight these. Only in the event that a CCD or other sophisticated form of imaging becomes part of your activities at the telescope will you need much more than the simplest finder, and a larger secondary guide scope required. It is not unusual for a telescope of 12 or more inches (30 cm) of aperture to feature a fine 4-in. (10 cm) (or even bigger!) refractor mounted piggyback along its tube. However, such astronomical activities certainly lie within the realm of most people reading this book.

Color filters? Only a maybe; be sure to read Chaps. 3,9,10 and 11.

What about imaging? If you want to join the craze, what would be the minimum you would need? If time is of the essence, before you go any further you should know that most advanced systems will probably require more time than you have. However, the desire to record something of what you have seen may well remain, so you might take the simplest and most personal route of all, and just get a good sketchbook, pencils, and erasers. You do not need anything more than this in order to have some of the most exciting adventures possible in astronomy.

It is only in the relatively recent past that electronic and other sophisticated imaging has been available as a viable option for the amateur; most of those long in the hobby never contemplated some of today's grand approaches. Standard digital cameras are hugely valuable and in the right circumstances can provide the fastest, and simplest, imaging method of all. However, as far as the purest joys of astronomy are concerned, one could argue that we have lost more than we have gained with many of the new devices. Certainly we have lost much in the development of advanced viewing skills that drawing imparts, and the cost of this simplest of ways is obviously minimal. It would thus seem a good idea to spend at least some time sketching at the eyepiece.

Besides drawing, there are also some quite inexpensive options in the marketplace today that will allow us to achieve some very good results fairly quickly and easily:

- *CCD video cameras* especially built for astronomy and which do not necessarily take too much of your time to learn how to use
- The simplest *CCD cameras*, which perform quite well but require more of your time
- The lowly *Web Cam*, which turns out to be amazingly capable of producing results completely out of line with its humble status, but requiring yet more time from you than you might be able to give.

These last three options, to a greater or lesser degree, are more time-consuming to use than the simplest approaches previously listed, but they may well strike just the right balance for your own circumstances. A big plus is that few of the aforementioned require large outlays of cash although, for fast deep space imaging, standard digital cameras will require the combined use of an image intensifier, always an expensive item.

For advanced CCD imaging and all that this entails, it is a different matter entirely, and for our purposes it is best left alone.

CHAPTER THREE

How to Expand Your Potential

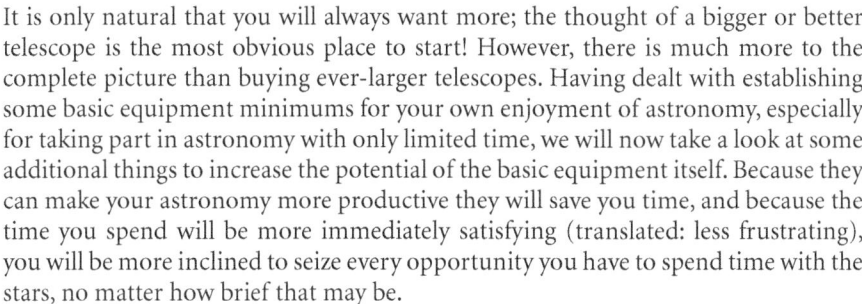

It is only natural that you will always want more; the thought of a bigger or better telescope is the most obvious place to start! However, there is much more to the complete picture than buying ever-larger telescopes. Having dealt with establishing some basic equipment minimums for your own enjoyment of astronomy, especially for taking part in astronomy with only limited time, we will now take a look at some additional things to increase the potential of the basic equipment itself. Because they can make your astronomy more productive they will save you time, and because the time you spend will be more immediately satisfying (translated: less frustrating), you will be more inclined to seize every opportunity you have to spend time with the stars, no matter how brief that may be.

Aside from the telescope (whether one to get you started or one that represents your final destination), plus some other key components, there are some pretty impressive accessories these days to tempt you further. Some of them actually have the effect of increasing the aperture of your telescope, and do, indeed, deliver on the promise. There are numerous other possibilities as well. Modern eyepieces, focusers, mountings, light pollution and narrowband transmission filters, digital setting circles, electric microfocusing, and even dual eyepiece stereo viewers also bring new potential to your telescope; certainly, there have never been more and better accessories to choose from. Without presenting a comprehensive survey of all that is available, here are some comments on those things that may best help you to attain some of the objectives of this book.

Comments in the last chapter regarding all things hi-tech need to be taken in the context in which they were intended, and not to discount the fact that some of the best accessories for our purposes also embrace the best of today's new technologies. Indeed, there is hardly even a conventional component for astronomy these days that does not now have the advantage of a more efficient and accurate method

A. Cooke, *Make Time for the Stars: Fitting Astronomy into Your Busy Life,*
DOI: 10.1007/978-0-387-89341-9_3, © Springer Science+Business Media, LLC 2009

of production; this is quite separate from the newest observing tools entering the marketplace. However, this does not mean that everything produced today is of good quality or design, or that everything from the past is necessarily inferior – far from it.

Most new directions in amateur astronomy require additional applications and expertise, possibly with a greater investment of time than you may have. This is because they are intrinsically locked to further tech applications; very few of them enable a direct or immediate approach. For us, therefore, perhaps the best are those that may be used without additional "layers" – not requiring, in themselves, that *we* somehow adapt to them! Of most interest to us are those that add to our existing activities, rather than steering us into some new avenue altogether. Such accessories, while not indispensable in order to take part in a meaningful and personal acquaintance with the stars, are all sufficiently valuable that you may want to consider them as your own priorities evolve.

The Best of the Bunch

Equatorial Tracking

One of the best ways to expand your potential is typically quite low-tech, but deserves some mention here. There is nothing quite like the luxury of a fine, equatorial tracking telescope. Not having to worry about the constant chase to keep your subject centered in the field of view is something that can only be fully appreciated by those who have spent time without such benefit! It means that your limited time at the eyepiece is *all* spent viewing.

Today, it is possible to have the means to accomplish this capability with virtually any design of telescope mounting. Even the simplest of them all, the Dobsonian, has some acceptable tracking options available now. The simplest and most ingenious, originally known as the "Poncet Platform," provides reliable tracking for sufficiently long periods to solve most observers' requirements. Many examples have been built successfully by amateurs, and descriptions of how they may be constructed by the enthusiast are widely available. There are numerous resources on such home-built design available on-line. Some sophisticated and well-built models are also available commercially.

Better still, but more costly, are computer-controlled tracking drives that supply movement to both axes, adjusting in tandem to create perfect tracking. This is now the standard form of tracking used in the largest observatory telescopes, with mountings closely related to the simple Dobsonian. It is also commonplace on many commercial catadioptrics. Although it is a complicated system, eliminating immediate viewing because of the requirement for preliminary star alignment, it works well in practice once set up. But you do have to set it up, which might become a hindrance when time is short, so be careful about assuming it is an easy solution for you. It will also produce "field rotation" of the image (something undesirable in imaging), but this may be controlled using a "field derotator" at the focuser and is not a problem for visual use. Whether it is possible to fit such an elaborate tracking arrangement to your particular altazimuth mounting (should you have one) depends on many factors, so check out what might work for it, especially before you buy.

Of course, that old standard, the "German equatorial," and the Newtonian telescopes that its devotees often use, remain close to the core of amateur astronomy; when well built to correct mechanical problems (an important qualifier, since poorly made amateur mountings can be terrible!), the design's splendid and solid fundamentals ensure that it will remain a dominant mounting style for the foreseeable future. It remains very common on commercially produced instruments, and for moderate-size reflectors it brings the intersection of the two axes close to the center of the telescope tube, ensuring that the eyepiece position is more likely to be favorably placed. However, with larger reflectors of even short focal ratios, problems of unwieldiness increase for the observer as the eyepiece position is elevated ever higher.

At the other end of the scale, with small reflectors we may find ourselves stooping; effective use of such instruments will necessitate using tall tripods or piers, something not necessarily conducive to stability or ease of use. With small to moderate refractors, the problem may be more severe still as we find ourselves stooping to look upward through the rear-mounted eyepiece position; even taller tripods or piers will be required, and the use of star diagonals do not necessarily eliminate the problem. Large refractors are ideal with this design of mounting, as the eyepiece is placed increasingly high on a central pedestal for comfortable viewing. Indeed, all of the great refractors of the late nineteenth century feature German equatorial mountings, although normally they are so enormous nevertheless as to require special elevated viewing stands for much of their range.

A visit to the pages once again of Volume 1 of *Amateur Telescope Making* will show many other equatorial mounting types. Many will simply take up too much room, or rule out portability, such as the old-fashioned yoke-type mounting. However, no one could dispute the effectiveness of any of these designs when solidly constructed. Others, such as a compact fork-type configuration for telescopes with short focal ratios and a wedge for equatorial tracking may well be highly suitable to your needs; there are many older catadioptric telescopes equipped with such mountings. These are basically just altazimuth mountings with the base set at an angle to place its axis in alignment with the pole. With just a little ingenuity, this type of dual axis full equatorial tracking may be applied to any altazimuth-mounted telescope.

Most of all the advantage of any equatorial design is simplicity of tracking, which will still work quite well even when only roughly aligned. Uniting all equatorials is the ability to slew the telescope in two principal directions at right angles to each other, when motorized the main benefit being to free the observer from the constant need to center and follow the object under scrutiny. Do they really save you time? They may well be among the most important time-saving items you can have.

Eyepieces

For developing our level of viewing refinement to extract the maximum results, purchasing better eyepieces should rank high on the list. Today's eyepiece designs and production methods have resulted in a huge improvement over anything we had in the past. There is a wide array to consider, so much so that you may find it hard to choose, or even to know where to begin! Here is a prime example of applied technology that seeks not to replace anything, but to improve upon all that has long been valid. With complex and sophisticated multilens designs having become commonplace, with previously unimagined optical performance, let us hope that the visual astronomer survives in sufficient numbers so that such magnificent eyepieces will continue to be made!

When Al Nagler (founder of TeleVue Optics and the original pioneer of these grand new designs) dared to introduce his first range of ultraspecial eyepieces into the marketplace, speculation was rife that the amateur community would not embrace them. Despite producing items with extraordinary specifications (not possible at any price only a few years before), the new eyepieces were still far from cheap! Critics said they were simply too expensive to be of interest to amateur observers, who might have paid less for their scopes than some of these eyepieces! In fact, they inhabit a price realm previously unheard of within amateur circles. As it turned out, the pundits were proven to be quite wrong; acceptance was rapid and widespread, and buyers were not deterred at all by the prices. The performance of what were often monster optical configurations produced not only breathtakingly wide, flat fields of view, but also sometimes even corrected inherent optical weaknesses in telescope designs, such as coma in Newtonians. Had the naysayers had their way, the live observing community would have been much the poorer (although we are indeed poorer, in money, having had to pay for them!).

The wide light cones of many modern larger, short focal length telescopes necessitated field lenses in these eyepieces of unprecedented widths, and thus many wide-field eyepieces were supplied in 2 in. (51 mm). It had also become important, in fact, to consider such wide light cones in the optical design of all telescopes, because the potential of the new eyepiece sizes would be lost on optical paths too narrow to utilize them, especially where secondary mirrors clipped the cone (vignetting). Thus, 2-in. focusing units and larger secondary mirrors became the norm in such telescopes, although such massive eyepieces were unheard of in amateur circles before Al Nagler.

However, the use of 2-in. eyepieces brought about an additional consideration: telescope tube balance. Makers and users were forced to consider this issue, especially with some of the larger assemblies, such as Dobsonians. The sheer size of the new eyepieces meant considerable added weight at the viewing end of the telescope! Pick up one of the larger eyepieces and you will no doubt be astounded at just how much it weighs; many weigh more than a pair of binoculars! You have probably picked up many small telescopes that weigh less! Some suppliers (such as Orion) began marketing special tension springing on their Dobsonians, which apparently produces excellent results in countering lopsided balance problems. Other types of mounting may necessitate strategically placed counterweights or tightening screws. If perfect balance is a necessity on your own telescope, you will need to add more counterweights low on the tube, which may not be the easiest thing to do, depending on its design. The use of smooth progressive clamping on both axes for tracking may make it unnecessary, even if exact balance is not achieved.

You may be asking yourself just how good are these grand new hi-tech eyepieces compared with what was available before? Well, you might want to save your pennies. They are so good that, once having used them, it will be hard to return to a lesser form, at least, willingly. Frankly, these complex, multilens designs put even that old "dream" eyepiece, the Erfle, to shame. And you probably know just how good Erfles are! Not so long ago, we used to dream about owning just *one* of these classics, usually ex-military adaptations that had been produced "without regard for cost"! We used to relish the thought of owning a fine set of orthoscopics as well for planetary viewing. Even a Kellner eyepiece seemed quite excellent. (Actually, it is not a bad eyepiece at all!) In truth, none of these are poor choices, even by standards nowadays. However, by comparison, it soon becomes readily apparent that for deep space viewing, spectacular panoramas of the Moon, or great high-power planetary performance with exceptional eye relief, there is nothing quite like the new designs. It is wonderful to take in the entire dimensions of an extended object, and see it with high transparency, flat field, and full color correction. The wide field varieties (such as "Panoptics" and "Naglers") have been compared to looking through a picture window in space. Such luxurious designs make all forms of viewing a new and relaxing delight. Many feature exceptional eye relief, which can make it very satisfying to view subjects at high magnifications without having to have one's eye almost touching the eye lens. Similarly, one need not tolerate the bright "ghosting" in planetary observing so familiar to observers of old; in the solar system the contrast and high light transmission of eyepieces such as TeleVue "Radians" put all the old designs to shame. "Radian" eyepieces also boast a special feature allowing you to adapt the eye relief setting specifically to that of your own eyes.

On the less costly but no less hi-tech side, there are also many less expensive eyepieces by many quality manufacturers. (Take a look at the range of eyepieces offered by Orion of Santa Cruz, California, just to mention one source.) The designs of these alternative eyepieces usually produce narrower fields of view than those by TeleVue (though, usually their fields of view are considerably wider than most of the old designs), and cost far less than the most glamorous wide field designs we have just covered. In a few cases, they may actually provide superior views to their grander, more costly cousins, at least in the central part of the field. In referring to superior views, we mean all those subtle realizations of color and contrast that planetary observers cherish so much. The reason for these qualities is not hard to grasp: simpler eyepieces utilize fewer internal lens elements, which may result in less light absorption and fewer possible internal reflections, at least in the best quality examples. Plössl designs are particularly effective in this regard and are often relatively inexpensive. Other designs related to Plössls often utilize the word "Plössl" as part of their name, such as "Super Plössl." If maximum contrast is your aim, just be sure to check the number of lens elements, or the performance in this respect may be less than in the original standard *Plössl* version.

Once again, in this instance there can be no question that traditional observing was handsomely served by technology, making it far more effective and enjoyable. As it turned out, Nagler's products had not only proved widely successful, they had become the benchmark for all other eyepieces by all other manufacturers since. However, one should not preclude the prospect of selecting fine eyepieces that may provide qualities or specifications other than what TeleVue currently offers, and even the real possibility of saving some money. Just be sure to do your homework so you know what you are getting. And you do not need a box full of them. Even just three good ones and a Barlow will streamline your time in ways you will soon appreciate for yourself.

Focusers

Although good focusing units have long been available, increasing needs for flexibility of use as well as precision in focusing necessitated more refined, even electrically powered, focusing mounts to match the new optically excellent and large eyepiece designs. The 2-in. (51 mm) sizes soon became almost standard. Because high-power eyepieces were unlikely to be made in the new larger 2-in. sizes, adapters became necessary for the focusers of 1¼-in. (31.75 mm) eyepiece sizes. Different breeds of telescope complicated the matter still further. Because of the differing eyepiece travel they required, catadioptrics, reflectors, and refractors required ever more sophistication and specialization for their focusing units, and the old "one size fits all" rack-and-pinion approach was clearly a thing of the past. The minute adjustments required in catadioptric telescopes make them particularly troublesome in this regard when supplied with only coarse focusing mounts.

Although the new focuser designs, of course, represented significant new advancements, they also meant yet further expense! However, who now, having tried them, would be satisfied with anything less? JMI were soon to become famous for their development of "zero image shift" Crayford focuser designs, which utilizes rollers rather than rack and pinion or helical gearing. This type of focuser, with its rock stable image positions and precision small adjustments, soon became the benchmark in the industry. Meanwhile, the continuing development of Crayford designs continues to this day, including digital readouts and microadjustments, although these improvements are aimed mostly at increased precision and control for advanced imaging.

Crayford units feature ultrafine, continuous, and smooth-moving increments of focus, with little or no movement of the subject in the field throughout the focuser's travel back and forth. They lends themselves well to electric remote controllers. Such remote electric focusing provides truly jiggle-free adjustments! However, there have been other refinements of the original concept of the Crayford to the present day by numerous manufacturers. Special mention should be made of Van Slyke Engineering, a company that makes the most extraordinarily beautiful and minutely precise (not to mention extraordinarily expensive!) focusers you might ever see. However, before getting carried away with great extra expense to have the new refinements and capabilities of such focusers as these, be sure to recognize what actually will be of value to you. You may find no reason to advance beyond the basic design utilized in most Crayford-equipped telescopes. So it comes down to what your own specific application is (just as in so many things astronomical these days), which is where your needs may part ways with fancier forms of focuser, since the greatest practical value of the more advanced models would only be realized in CCD imaging.

However, the old helical twist-type focuser still continues to hold certain value for some Newtonian reflector users, namely simplicity and complete fractional focusing accuracy, even if motorization of focus is not readily available for them. So, do you really *need* electric focusing? Not exactly, but you will find having it adds very considerably to your pleasure as well as the satisfaction of observing; the absence of constant focusing jiggles, along with the ease of making fine adjustments, likely will make it more of a necessity in your own personal list. These once-luxury features do not even represent large outlays of cash.

Light Pollution Filters

In a rather stark comparison to the comments regarding conventional filters (see Chap. 9), light pollution filters are worthy of high praise. Maybe you already know the benefits they offer. Considering the huge "bang for the buck" they provide, there can be no better, more useful, or more powerful accessory. A different kind of beast from the standard color filter, they literally shut out certain wavelengths from the visible light spectrum and transmit others, and to great effect at that. Even in their physical appearance, these filters are strikingly different to any other, in hue, from front to back, and in their apparent high surface reflectivity. It is immediately clear that these are not merely simple color filters! Best of all, not only are they particularly valuable in suburban environments, but in dark sky conditions they can enhance our views amazingly as well. Many are designed to respond to selected wavelengths of light originating from ionized hydrogen gas, which is widespread throughout the cosmos, and filter out other wavelengths (especially man-made!), but you will have to experiment for yourself. Find the right object, and it can actually seem to light up against the background sky with new luminescence and detail; at its best, it throws chosen subjects into dramatic contrast against the sky. However, as with everything, none of the available filters will be equally valuable for all things, which is why many observers favor having a collection of them!

Although many types exist, each of which has specially "tuned" wavelength transmitting and blocking properties, overall, the narrowband varieties are generally most useful, as they give favorable results on far more subjects. Broadband filters would seem to be of less value, although they certainly improve the effects of a poor sky. Specialized forms of narrowband filter for select uses include Lumicon's former product range (now part of the Parks Optical Company and catalog). These have been well known for many years; perhaps the most celebrated filter in this line is made specifically for viewing just the "Horsehead Nebula," a tricky object at the best of times. Among general-purpose narrowband filters, Orion's Ultrablock filter still rates at top of the heap, although there are certainly numerous products almost as good, with similar characteristics, by different manufacturers.

Although light pollution filters may seem expensive compared with regular color filters, they offer far greater returns. You could buy two or three for the cost of one medium-price TeleVue eyepiece, so it is not too great a financial plunge to take, despite such relative cost. And they will allow you to see much more from your home base – likely to be the busy astronomer's main observing location.

For filters in general, you might refer to Chap. 9, as well as the specific opinions featured on the website: (http://sciastro.net/porta/advice/filters.htm).

This site is quite instructive for comments and guidance regarding many filters of all types. There is a wealth of detail on the use of many conventional varieties, and you may find objective contrast to this author's sentiments on the subject. However, remember not to invest in quantities of regular color filters until you have firsthand experience with just a few. It is important to have keen awareness of just what you should expect them to do for you.

Image Intensifiers

Image intensifiers continue to evolve to higher forms and should be considered seriously by anyone looking to maximize deep space viewing. They also allow very satisfactory viewing from less than ideal locations, which are likely to be where we do most of our viewing, so their value cannot be overstressed. The controversy surrounding these amazing devices has still to die down, but suffice it to say, they are so remarkable, and the advancements recently made in them so significant that to ignore the topic would be performing a grave disservice. While this volume is not primarily concerned with their use, you should have at least the latest information so that you can make up your own mind about them. Perhaps needless to say, these devices are not for the more brilliant objects of the solar system; think of them primarily as deep space equipment.

Recently Collins Electro Optics (the only company in the world offering specific image intensifier products to the astronomer) began marketing an even higher performance version of their original image intensifier eyepiece, utilizing what was previously referred to sometimes as a "Pinnacle" tube. These tubes are made by ITT and utilized in the I3 unit. Although they do have quite an edge on the competition, placing them at the top of the pack, it does not negate the value of other manufacturers' tubes. Do not despair if you cannot obtain a Collins I3 or ITT-based device! Other tubes by different manufacturers and with similar advanced designations of generation (Generation 4) are available in the USA and also overseas, and will probably be more valuable than many Generation 3 versions. Thus, you should try to acquire the most advanced intensifier technology available to you, as every incremental degree of refinement does indeed make a difference. The slightly ingenious enthusiast may readily build an image intensifier eyepiece; the tube acts as the field lens, and a simple Plössl eyepiece may be attached to the other end to magnify and flatten the effect of the tiny concave phosphor screen.

The newest image intensifier generation had previously experienced some confusion over its proper designation. So-called Generation 4 tubes were originally considered to be Generation 3 tubes, only "*with thin film technology.*" No official consensus seemed to exist until recently that this more advanced system was in fact a completely new generation. However, the effect it brought certainly made very clear that it was! So now, apparently it is legitimate to designate *thin film technology* tubes as Generation 4; regardless, no matter what you call them, you now know what they actually are. The thin film Generation 4 version not only produces even less signal-to-noise ratio than the original standard Generation 3 model (resulting in surprisingly better contrast), but also increases the visual output significantly. In fact, depending on your circumstances, the boost may even seem to be by as much as 300%! Only relatively recently have the new versions been offered by Collins as standard issue, and at little extra cost.

Early assessments of the potential of these more advanced nonmilitary Generation 4 tubes greatly underestimated the improvement that is actually possible. At the time it seemed reasonable that the extra cost, at nearly twice the price of the standard Generation 3 unit, would not justify the gain in performance. The cost of a Generation 4 tube now may be only slightly higher than the old Generation 3 system, depending on manufacturer. Because of the new pricing, and the greatly superior performance it offers, you should try to acquire this more advanced intensifier generation. As a further point of information, the improved resolution of the new tube is also markedly superior to *any* other commercially available visual electronic device, and this includes today's very popular frame integrating video cameras. These cameras are always limited by the number of lines utilized by video systems, and they simply cannot provide the fine resolution of the new intensifier eyepiece, let alone recreate the same refinement of image that the advanced intensifier tube is able to produce. But most importantly, no video system can produce such resolution and "living presence" in true real-time, and by just peering into an eyepiece at that!

The old Generation 3 system previously provided a boost of 2–3 times the aperture used. (I estimated that my 18-in. telescope had been performing with an equivalent light grasp of something approximating a 50-in. telescope.) The new intensifier eyepiece offers about twice the gain we had before. (Now, the same telescope is apparently performing with the light grasp of more akin to the Mount Wilson 100 in., and possibly even closer to the Mount Palomar 200 in.!) From home, that most elusive and distant globular, M54, was split wide open into stellar components, almost to the core! Not bad for an object usually described in appearance as merely "granular" around the edges through even the largest amateur scopes. The same night, the much increased light grasp was also very apparent for such old favorites as the edge-on galaxy NGC 5866, whose tiny dust belt and considerable mottling was exquisitely detailed, refined, and revealed. Acceptable digital snapshots of many objects could be made with little difficulty (with never much more than a second or two exposure). Although the light-polluted suburban background sky takes on the predominantly green color of the phosphor tube, the eye soon learns to ignore it. The image below corresponded quite closely to the appearance of the suburban telescopic views as seen through the image intensifier eyepiece itself. It was imaged with a rather unsuitable, relatively low-resolution camera (3.2 megapixels), with no special settings; even so, the image is remarkably good (Fig. 3.1).

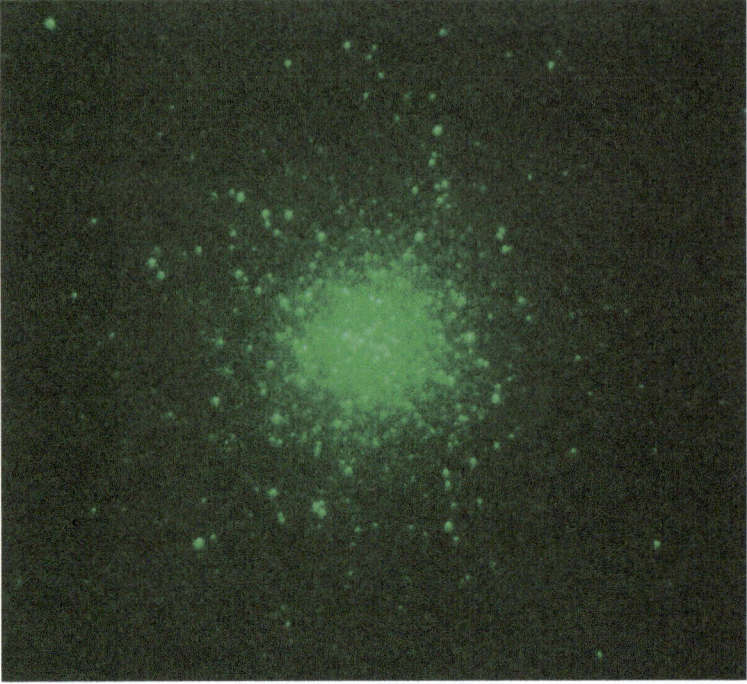

Fig. 3.1. The Hercules Cluster M13. (An 18-in. telescope with Collins *I3* Generation IV image intensifier eyepiece in heavily light polluted suburban skies. Minolta Dimage digital camera; 3.2 megapixels, exposure 1 s.).

M13 features spiral chains of stars around its circumference, giving it the appearance of a cosmic spider, better seen in conventional viewing than in the image here. Lord Rosse's famous, or once infamous, "Propeller Lanes" have received much attention in recent years, since they were at one time believed to have vanished. The lanes are situated on this image at the 2 o'clock position, and you will see the hallmark triple-armed cross that gives the legendary dark lanes their name. Spend more than a superficial look at the cluster, and the lanes begin to make themselves known, and then it seems that they dominate what you see! They are even easier to see with an image intensifier. The fact that they had not disappeared after all has always been a source of intrigue, since they are clearly visible on many old photographic plates. Perhaps observers simply forgot where to look (they are to the side of the cluster's center), or maybe they did not stand out in the way many people had expected. However, they are there to see, and in a highly visible form at that, although not always immediately apparent at first glance.

At a dark sky site in the desert, the new image intensifier was astounding: the quality of enhancement of M2 had a brilliance, resolution, and refinement impossible to describe; the Ring Nebula M57 looked even better than many photographs from great observatories. With a sizable amateur scope, imagine seeing the "ansae" themselves on the "Saturn Nebula" NGC 7009 in real time; actually, you do not have to imagine, you can see them. No matter how good the views through the previous Generation 3 model were, by no means can they compete with the much brighter, essentially photographic quality of the images that glow in the field of view of the new device. Most significantly, while never overcoming the negative aspects of suburban viewing, the views through the latest intensifier actually are much closer to the way things appeared with the old intensifier under dark skies.

Sadly, the great value of image intensified astronomy is still waiting to be discovered by the average amateur observer, and even more unfortunately, the Collins I3 remains alone, even relatively unrecognized in the marketplace. Many astronomy forums and blogs have all kinds of comments made by supposedly insightful folks who, in actuality, have no idea what they are talking about, especially concerning *any* modern image intensifier system. Add to this that intensifiers are likely to remain the most expensive live viewing accessory of all; worse, the particular model you want is only available in the USA at this time (regulations mean the most advanced versions may not be exported overseas), and you can see that they cannot be for everyone. Perhaps many people feel it is easier to knock them down than build them up. If you are unable to obtain a Collins unit, either because of country or cost, and you wish to try image-intensified astronomy for yourself, there are nevertheless some other viable options. As long as you are prepared to do a certain amount of fabrication and assembling of the various components for yourself the potential is tremendous. (see *Visual Astronomy Under Dark Skies,* Springer 2005.) Additionally, Collins can supply complete housings of the I3, everything including the optical components minus the intensifier tube, in which certain other intensifier tubes, even from outside the USA, may be fitted.

It seems also quite significant that inexperienced observers are more likely to be able to see details described by more experienced observers when using these devices. There is far less of a learning curve. And just as another added bonus to image intensifiers in general: if you are at all interested in artificial satellites, you will never be more aware of them when you are observing than when you are viewing through an image intensifier! Hardly a minute goes by without a satellite crisscrossing the field of view, in an amazing demonstration of just how many man-made objects there are whizzing around in Earth orbit! A sobering realization, indeed. So while image intensifiers are hardly necessities in the purest sense, for those who have become their devotees it certainly seems that they are. If time is not your ally, and you cannot often travel to remote sites, these devices will allow you to see many things in deep space with much of the glory usually reserved for conventional viewing from far darker locations.

Binocular Viewers

Because our eyes were designed to use together, over the years, many amateur builders, and even a few professional manufacturers (such as JMI, who introduced their own line of reflecting binocular telescopes in recent years) have produced binocular-style double telescopes. The views must be extraordinary. Once smitten, some people have found these types of telescopes to be the only way to go, despite the considerable expense of matching two complete and separate identical optical trains in tandem, and the numerous difficulties of holding these two trains in perfect mechanical alignment in all mechanical positions. There can be no allowable flexure between the two! For the rest of us, with our mere *single* aperture scopes(!), there is still a way to access most of the benefits of true stereo viewing. Binocular viewers for single apertures are becoming more widely available, and they may change your definition on what constitutes, or does not, an essential accessory.

Since the benefits of parallax on two eyes do not exist as we look far into infinity, stereo vision is not possible by the ordinarily accepted explanation, even for a true binocular telescope. However, somehow a magical sense of depth perception is still present in both of these double-eye systems. With certain technical differences, this dimensional awareness is happening in much the same way, so the mechanism for perceiving it in the brain must also be the same. According to the most commonly accepted theories, we should not actually be experiencing three-dimensional sight in the absence of parallax, and yet to some degree we still do. Standard explanations of three-dimensional vision describe the slightly differing images of each eye being compared in the optical center of the brain. In reality, the *totality* of our depth perception must be attained by more than this way alone (presumably via the width between our eyes themselves, and some other effects of the magnified and compressed focal plane), because the brain still is able to impose some stereo processing, even though with a binocular viewer we are only utilizing a *single* aperture! The dimensional effect of the Moon is startling, and even the planets seem to take on certain three-dimensional qualities. Deep space itself seems to take on more depth, and although it cannot be claimed that we experience all of the magnificence of a true three-dimensional view, we certainly see more than a merely flat image. Star clusters seem to make the brightest stars appear to be nearer, giving the clusters a more globe-like appearance, and nebulae and galaxies seem to float amidst the void, revealing more subtleties and "folds" in their structure. Overall, the effect is remarkable and still does not seem not fully explainable by any theory!

Additionally, more resolution also becomes possible in stereo, probably partly because we are so much more relaxed at the eyepiece; it is so striking that you may never go back to the single-eyed approach! If the potential for such visual luxury sounds tantalizing, you should seriously consider acquiring a stereo viewer, but before diving in you may need some background on these accessories, because using one is not as simple as it sounds. To begin with, they require the use of two identical eyepieces at a time, which will certainly increase the cost, and so it is not just the binocular device itself that you will need but sets of *truly* identical eyepieces, too. However, various *single* Barlow lenses of different powers can be of great service in this respect by providing more range of power for whatever matched eyepieces you have, since they work on the "front end" of the unit.

You must also keep in mind that you are splitting a single light beam into two parts, 50% for each eye. Since the brain is already accustomed to a single image from one eye, seeing an image instead via two eyes is quite possibly why the 50% reduction in light per eye is not perceived to be as much as this. It would appear that the brain merges the light sensitivity of both eyes to make a whole at 100%; perhaps this is not a scientific explanation, but it does at least seem a logical one! In any event, the great benefits through virtually any telescope should, in most instances, more than make up for the slight loss in brightness. Actually, once these viewers are coupled to apertures beyond moderate sizes, the downsides of splitting light two ways within a single system become increasingly hard to detect.

One of the greatest unexpected bonuses is in viewing the Moon. Previously, when using moderate or low magnifications, the brilliance of it in larger tele-scopes is completely overwhelming, dazzling your eye into temporary blindness! Lunar filters became necessary therefore for most observers, but do not provide the most pleasing view. Somehow, the effect is one of dulled vision, even though the resolution remains the same. Remarkably, the split light path of the binocular viewer essentially eliminates this problem, and the lunar filter is rendered unnec-essary. The reduction in brightness is not noticeable, except in the benefit to your eyes! There are other great benefits of binocular viewers as well, which go a long way in offsetting any potential for decreased image brightness. These include the ability to make out increased contrast, as well as an increased sensitivity to low light levels. Even if you perceive decreased overall brightness, surprisingly, viewing through both eyes can result in the ability to detect otherwise unseen objects, and improved detection of structural form in otherwise vague objects. You will also experience less annoying interference of "floaters," so typical in higher magnifica-tions during single eye viewing.

For obvious reasons (cost!!), it is unlikely that you will acquire two image inten-sifier eyepieces. However, if you ever have the chance to view through two of them at once, it is an opportunity not to be missed, the ultimate form of deep space live viewing! It works well even when using an older original intensifier along with the new version. Old and new Collins intensifiers are identical optically, using the same matched TeleVue 25-mm (1-in.) eye lens component, and the two image intensi-fiers work dazzlingly in tandem. In the total visual equation the superior image shown (set to your dominant eye) in the newer unit wins; the lesser intensifier fills in the rest of the stereo impression. Even the somewhat increased magnification (from the 1.6× Barlow lens attachment) seems often to produce a more ideal size for many of the grander objects in deep space. Fainter subjects do increase elec-tronic "noise" to less than ideal acceptable levels, so you should reserve single eye vision for them. This, of course, points to the more limited magnification range you will experience with image intensifiers in general, so you must select powers wisely.

Typical views with a single intensifier appear visually "flat," as is the norm with these devices. Utilizing two together, even the depths of space itself, suddenly take on a depth of dimension that was lacking before. Of course, there is the bonus of the easy and comfortable viewing with both eyes at once, but who would have imagined ever doing it in *this* extravagant way?! There is a further benefit for smaller apertures too: the resulting reduction in total brightness of the divided optical path is less noticeable than in conventional viewing. The visible output of an image intensifier is more contingent on the brightness of the object we are studying than other factors.

One thing you may not be prepared for could be the sheer weight of the viewer with one, let alone two intensifier eyepieces! It will certainly cause balance and other problems in many instruments. For Newtonians with rotating tube sections, a simple friction clutch may provide the answer. Made from readily available materials from a local hardware store (some stout aluminum angle, a twist-screw deadbolt assembly, and a rubber stopper), it is joined to the controller bracket attached to the rotating upper portion. This simply clamps this part of the tube to the nonrotating tube portion in any desired position.

Before buying the first binocular viewer that comes along, you must resolve certain issues. Although theoretically these viewers may be used for any type of telescope and eyepiece, in practice, you should select carefully, because they are all just a little different in ways that will directly affect their compatibility with your particular scope. There are also two distinct types of viewer. Do not choose the older, angled (like a dual microscope viewer) type, which resembles the attitude of a 45° star diagonal, and optically, is also not the best design for telescopes. These viewers usually feature relatively small internal prisms and are unsuitable for lower powers utilizing eyepieces of moderate to longer focal lengths, where substantial vignetting may occur. Additionally there are focus issues as one adjusts the width of the viewer between eyepieces for different eyes. Far preferable is the straight through, true binocular prismatic style viewer, which allows the greatest flexibility of use with virtually all types of telescope. Having a design that allows full size prisms, it accommodates a far wider range of eyepieces and hence magnifications. Plus, just as with a standard binocular, it may be adjusted for varying distances between the user's eyes, has at least one adjustable eyepiece focus, and usually transmits more light than the angled microscope-type viewer. However, whichever viewer you use will certainly create difficulties in accommodating more than one observer at a time, because everyone's eyes are so different. Astronomical stereo vision is not nearly so simple as direct conventional viewing. However, for one user it is glorious!

Here are the purely mechanical considerations for stereo viewers:

1. The addition made to the total length of the optical path complicates things right from the outset. This is not only because of the dimensions of the viewer itself (which places eyepieces further away from the usual position of focus) but also the increased distance that light must now travel through the internal prisms. For many types of telescopes, especially Newtonians (with the typical short latitudes of their focusers), to achieve focus will simply prove too much without resorting to some corrective measures. Catadioptrics and refractors, having greater relative focal travel inside the focal plane, tend to fare better than Newtonians in this regard and will often not need such steps; most of the binocular viewers were probably designed with these types of telescopes in mind. And without a Barlow in certain telescopes of long focal ratios, stereo viewers may play havoc with these as well.

   For Newtonians, your lowest power Barlow lens will take care of most focusing problems, but consequently the image brightness will be diluted by the resulting higher power. But it does not stop there. Because of the greater distance now separating your eyepieces from those originally designed for a standard Barlow's use, amplification will probably be more than stated, maybe far more, further diluting the brightness! Luckily, certain makers of binocular viewer have made efforts to accommodate the problem and provide the means to bring the image into focus, with additional optical components, such as an internal and removable Barlow lens component (frequently of quite modest total power boosts, such as 1.3× or 1.6×) and built to provide only *that* particular power within the optical path of the viewer itself. It is because these Barlows are integral to the design of the binocular viewer that the stated magnifying power is what we can expect to see, as they provide just the additional boost needed to achieve focus, and no more.

There is another way to proceed, although it is considerably more cumbersome and inconvenient, and that is to alter the length of the telescope tube itself to accommodate the new focal position. It is possible to substitute shorter truss rods (or a shorter main tube, maybe by a removable extension) instead, but this seems a highly awkward solution for general use. Reconfiguring the telescope each time, depending on how the telescope is being used, is cumbersome enough to discourage using the telescope. In addition to this, shorter truss rods (or tubes) are not a completely satisfactory remedy, because a larger secondary mirror will now likely be required as well, creating a new problem.

2. The eye relief of long focal length eyepieces tends to be too much for two eyes to incorporate the entire widths of both fields of view simultaneously. More problematic is easily merging the images together. And very short focal lengths produce similar problems for the exact opposite reason! You will also find that the point of failure to merge the image with given eyepieces varies considerably from person to person. However, within a reasonable range of eyepiece focal lengths, the views will generally be successful for most observers and subjects, with inexperienced observers commenting on greater ease of seeing than with single eyepieces. This is especially the case for appreciating finer details often lost on their novice eyes. For planetary viewing, one can experiment with different additional Barlow lenses, although one as powerful as a TeleVue 5X Powermate will likely be a bit extreme, because far more than the stated power of 5× is the final result of the binocular viewer's optical configuration!

3. Some binocular viewers, when used with large eyepieces, or even two image intensifiers, impose a limit, by default, on how closely the two units may be brought together, simply because of the large diameters of each unit (2 in./5 cm). However, because of the way the image is formed in the optical train of intensifier eyepieces, the eye relief in them is unique; one may see satisfactorily from a wide range of distances from the eye lens, which helps. Thus, the intensifiers may actually be set quite close, since the need to position one's nose between them is less critical!

4. Should you also be fortunate enough to have the luxury of double intensifier viewing, bear this in mind: because the light train does not follow a direct, straight through path, such as in a conventional eyepiece, you may find that the two phosphor screen images do not precisely align and merge, no matter how you adjust the viewer. You may be seeing slightly different fields of view in each. There is a simple solution. Simply rotate one intensifier in its holder, and you will attain exact image alignment between the pair as the features common to the two fields converge exactly!

5. Although conventional eyepieces may be focused separately for each eye in many units, image intensifiers require that each unit be focused within itself. This means

that achieving focus is a two-step affair: first of all the telescope focuser must be adjusted to the closest focus possible on the phosphor screen of the intensifier; secondarily, the eye lens focus is then set to view that image in sharp focus. This component of each eye's optical path fulfills the function instead of the individual eyepiece focusers of the binocular viewer. Therefore, the first step is a factor of the telescope's own focus, and not the individual's eye.

Until recently, owning binocular viewers had always been a prohibitively expensive proposition, especially since they require the additional purchase of matching eyepieces. Are they worth the cost? There is good news these days; with the advent of some excellent new inexpensive viewers (such as those by Celestron and William Optics, both under $200), you may finally be lured on board the binocular viewer bandwagon. The William Optics version, for money and flexibility, seems to offer more than anything else in its price range. In fact, what it offers is quite remarkable by any standards, both in its precision and presentation. Aside from its high-end BaK-4 FMC prisms (fully multicoated best glass quality, for excellent light transmission), it is supplied with the added bonuses of two nice 20-mm eyepieces (!), an internal low power 1.6× Barlow component, and a wide range of individual focusing adjustments for each eye. The sophistication and very high quality of this particular binocular viewer belies its price, and it works extremely well. The eyepieces supplied certainly provide grand and relaxed vistas, and presumably were chosen because they are the best combination of overall focal length and width of field for conventional viewing, given their 66° apparent fields. There are traces of internal reflections and the slightest of spider diffraction effect on the brightest subjects, but these are very minor issues and common in these stereo viewing devices. Although there are numerous other binocular viewers available, some with increased optical flexibility, many cost up to 5 or 6 times more.

It is hard to claim that that stereo viewing will save you time in the direct sense. However, it does allow a much more relaxed and productive time at the eyepiece, again, all-important parts of the equation.

CCD Video Cameras

Although not really live viewing devices, CCD video cameras come pretty close. Those with frame integrating capabilities provide near live imaging and group viewing of deep space objects. Most users would probably consider this feature to be a necessity for video astronomy these days. Performing somewhat differently to image intensifiers regarding their response to the range of normally visible wavelengths and limits of resolution in video, many similarities still exist. They are often more sensitive than image intensifiers in blue wavelengths, whereas the most advanced generations of intensifiers are more especially suited to red and infrared. Each type of enhancing device's strongest suits depends on a specific object's spectrum, and the specifications of whatever unit is being used. Thus, the nature of the object may be revealed somewhat differently, and sometimes very impressively, although this is not always the case. It appears that not all deep space objects are created equal. And once in a while conventional viewing sometimes still wins in all respects!

For imaging, some quite advanced processing and compounding techniques may be utilized from saved video frames for some truly exceptional results, should imaging be your calling. Meanwhile, the best *nonintegrating* cameras still come into their own with certain solar system subjects. The older Astrovid 2000 – the original camera to create the CCD video boom – still works to great effect in this realm, a camera that nevertheless utilizes one of the most advanced CCD chips available. In fact, this chip is the same one used in the StellaCam EX! As a camera without frame integration, the 2000 will respond effectively only with brighter sources, such as those in the solar system, although what we are seeing is in true real-time. For viewing or imaging deep space subjects, or the faintest solar system members, the camera must be combined with an image intensifier, and you must understand its limitations.

Perhaps the best examples of CCD video cameras presently on the market are to be found in the StellaCam series from Adirondack Video Astronomy, which allow for up to 256 individual video frames to be integrated into a whole. This is twice the frame integration of the original version, and twice the capability of other competing cameras. The latest StellaCam models feature a better response in the red and infrared than the original, and though to a lesser degree, share something in common with the latest generations of image intensifier tubes. These cameras' refinement continues to grow, although each new version requires just a little more user expertise and effort than did the previous one. Today the latest camera in the line is the StellaCam III, which may be even more impressive than the SellaCam II.

A great advantage, which any of the cameras offer, is in allowing comfortable viewing with *both* eyes on a monitor screen. This relaxed approach sometimes actually makes it easier to see what is there in the first place and can serve as an excellent support system for the visual observer, something once again reminding us of the value of binocular viewers. We also have the potential of considerable image scale with these cameras. Within the solar system, at least, these cameras are capable of revealing large scale, fine, and remarkably well-resolved detail, but only as much as the camera's CCD chip and lines of resolution will allow. Of course, true real-time usage on solar system objects may be obtained by selecting only the nonintegrating function with an integrating CCD video camera.

All too often users do not understand how to get the best results from these cameras. First, we have to know at least something about the way they work. At the heart of each camera is a highly sensitive CCD chip. But regardless of its sensitivity, you must remember that the chip has a *finite number of pixels,* so these are not limitless, or even overly abundant! The best pure CCD cameras have the largest chips, and hence more total pixels; this is how they are able to provide such incredibly detailed views. However, even the chip inside the best CCD *video* camera has far fewer pixels than these. You therefore must try not to waste any of them on identical image information, because you have only a finite number available to produce an image. If you waste too many of them on too little information, much potential intricate detail is lost by their underutilization. Pixels are put to work fully only when the *minimum necessary are used to capture any given feature.* The effective focal ratio of any telescope should therefore be set up on the larger end of the scale to fully saturate the detail relative to pixels. We also have to establish a careful balance of such ideal saturation and what the atmospheric conditions of the time allow. As it turns out, on a good night, what might seem instinctively to be a far larger image scale on the monitor than your telescope is capable of supporting will more likely provide you best results for recording detail. Equivalent focal ratios of as much as F30, F40, and even more can be employed successfully, depending on the conditions and specifics of the telescope being used, and it is only at such scales that you will begin to capitalize on the camera's potential for resolving and capturing the finest details. There is nothing quite like seeing Mars appear as large as a tennis ball on your monitor screen!

Because the resolution of these cameras is also limited by the lines of the video system itself, once again larger image scales are the best way to capitalize on what you have. Otherwise, any one of the 600 lines of resolution of a camera may be smaller than the fine detail itself, meaning that more lines are used than necessary. Again, the wasted potential stems from the same basic reason. Although the state of the air ultimately dictates everything, it is always prudent to weigh all the factors that add up to the best possible theoretical use of your tools. A telescope with a focal length of around 72–84 in. (180–210 cm), combined with a TeleVue 5X Powermate Barlow lens, produces quite incredible views at an impressive scale. This works astoundingly in most atmospheric conditions, but is still *under*utilizing the potential of the video system itself! So, very occasionally, but only when seeing permits, you might combine one Barlow with another! Using 5× and 2× Barlows together would now enable total power boost of 10×, resulting in an incredible F45, more or less the optimal theoretical ratio! Planetary and lunar images at such powers on the monitor can be awesome. (Many Barlow lenses, such as a 1¼-in. and a giant 2-in. lens, are easily coupled together since one lens slips into the other.) However, use such extravagant powers sparingly and wisely! As you know, magnification can be a mixed blessing. Aside from the object we are observing, we also magnify vibration, tracking errors, and so forth, diluting the brightness and contrast of the image as we do so. Therefore, it is not always practical to utilize the maximum potential of the system.

Speaking of Barlow lenses, only buy the best achromatic varieties; anything else will degrade the images sufficiently to invalidate or reduce their value. Yet another way to produce adequate image scale would be to utilize that old standby, eyepiece projection, although this is considerably more cumbersome. You will need to experiment with your own equipment and needs, but always bear in mind the focal ratio potential as a guide, although it is admittedly not set in stone.

Light pollution filters, image intensifiers, and frame-integrating CCD video cameras can provide more illuminated or contrasted views of subjects otherwise too faint for satisfying astronomy from poor locations. For those who need to stay close to home to take part in whatever sessions they can muster at the telescope, any of these accessories are a godsend. This was much of the impetus behind *Visual Astronomy in the Suburbs* (Springer 2003), and these devices provided much of it. The crossover to the subject in mind for this book is never more apparent than with these devices.

Comparing CCD Video Cameras and Image Intensifiers

Some of the harshest critics of video astronomy consider that no form of electronic viewing has anything to do with live viewing, because these systems require that the observer use some type of electronic screen. Other critics have been even louder and more strident when it comes to the subject of image intensifiers, even modern varieties, which they may never have used. These are likely to be the very people who complain about the cost of these devices while routinely spending much more on other astronomical gadgetry, especially CDD imaging equipment! The cost seems never too much as long as we are *not* talking about attempts to expand the potential and range of live viewing! Furthermore, these same people seem to strongly advocate CCD imaging as the cornerstone of modern astronomy.

Although it is true that any electronic accessories take a certain amount of adjustment from an observer's perspective, this is also true of most things in conventional live viewing. Learning to see through a telescope is not easy, however we go about it, and acquired viewing skills are always at the forefront of requirements. However, it seems that a closed mind is hard to open, even when there is much to gain. At least CCD video cameras have been embraced by some of the amateur community. Unfortunately, the considerable virtues of these cameras are sometimes used to further denigrate the value of image intensifiers! Considering the widespread acclaim that frame integrating CCD video cameras had following their introduction, even greater praise for the significantly more expensive image intensifiers may be less likely to attract buyers, especially when one takes into account how well the video cameras perform. Maybe part of the reason for CCD video camera's acceptance is that actually they are a form of imaging device, for which much of the amateur community apparently has found a strange comfort level! Anyone who has used the best of today's intensifier technology knows how inappropriate a basis for comparison this is.

Certainly CCD video cameras offer good, relatively affordable, "near live" viewing, via the telescope. But while we apply these cameras to deep space, we should understand more clearly what they really do for us. The simple fact of the matter is that we must settle for something less than true real-time viewing; this is where the benefits as well as the drawbacks of frame integration come in. Although frame integrating models mimic the results of the intensifier quite well, by building up multiple frames of faint objects on the monitor screen to produce at least the effect of a live view, remember, these devices only *simulate* real-time viewing, and indirectly at that, on a monitor. On the plus side, although we may experience stunning views of deep space subjects, and in an apparent realization of real-time viewing, the downside is that what we are seeing is attained by accumulating the image over a period of time. The image we see may have taken several seconds to build, and in being seen indirectly on a monitor, the sense of direct connection to the object is lost. The unique experience of true real-time viewing is only obtainable when it is truly that – viewing through an eyepiece. Unlike the straight-through simplicity and ease of using image intensifier eyepieces, video cameras require a monitor, control boxes, wiring, a power source, etc., which ultimately adds to the sense of separation between object and viewer. This, then, is the price we must pay for the relative affordability they offer. Although it does not negate the considerable value of the video camera, it further illustrates the uniqueness of the more costly (regrettably!) image intensifier. Image intensifiers and CCD video cameras have their own spectral sensitive characteristics, and although results may differ, it is only in the details and not overall.

While all types of electronic devices will greatly add to your observing experience, because of the fact that they are primarily *enhancing* devices, it would nevertheless be foolhardy to imply that you actually *need* them. All of the various electronic aids have their own particular strengths and weaknesses, and their very expense (especially image intensifiers) may well eliminate them as potential options for many enthusiasts. However, if you want to give them a go, do not be off put by those who have nothing positive to say about them. The big plus is that any of these new electronic aids will effectively boost viewing brightness of otherwise faint deep space subjects. So, is the price difference still worth it for the image intensifier? Yes, and by every penny, if you can get your hands on one. Further discussion and comparisons can be found in *Visual Astronomy Under Dark Skies.*

A Word on Private Observatories

It is probably every budding astronomer's dream to have his or her own observatory. It is not a necessity, by any means, but just the thought of being comfortably housed in the dark hours of the night, your equipment "permanently" installed, always ready to go into action at the drop of a hat, no set-up time ever needed; just open the dome and away you go…. Would not that be fantastic?

Unfortunately, for most people this has turned into just another pipe dream. It is unrealistic because there are only so many sights these days that will accommodate live viewing of a quality sufficient to justify such a structure. The sad truth is that our cities have become huge light basins, which effectively eliminate the best of the night sky for most observers. However, if you almost never relocate to dark sky country, what is the downside of having an observatory-housed instrument ready to go at all times? Aside from cost and the space to put the observatory, this might be just what you need to make astronomy a reality for you.

For most situations, you may come to accept a compromise of sorts. If you are fortunate in being able to access dark skies with some regularity, it makes little sense to increase the challenge of disassembling and moving a bulky scope, having set it up semipermanently in a small building or under a roll-off roof-type observatory. And since many telescopes are built with portability in mind, it is fairly easy to move them the short distance from your house to the outside. You may have other ideas.

And Finally...

Location remains one of the biggest factors, no matter what equipment you have at your disposal. One always has to remember that while any quality enhancing device can indeed produce impressive results (and in some cases, startling results) in environments polluted by light or disturbed air, there are nevertheless upper limits on what may be reasonably expected of them in such places. (H-alpha filters have been reported to enhance the views through intensifiers and CCD video cameras in suburban locales.) So while the new electronic aids may greatly expand the potential of your viewing at home, depending on the nature of the object being viewed, they really come into their own in truly dark and transparent air, the ideal situation in which to take advantage of them. You will be freshly "knocked off your feet" by what can be seen in the best conditions at such remote sites, in spite of probably having experienced the same wonder many times in the past!

Ultimately, however, it does not matter which form of viewing you settle on, because under dark skies, the possibilities seem almost limitless, regardless. So, try to take advantage of any chance to get to these places. Although enhanced viewing does change the ease and visual drama, it can never replace the pristine beauty of traditional natural viewing, which remains the simplest and cheapest option of all.

CHAPTER FOUR

Maximizing Your Time at the Telescope

Now for the final ingredient of the equipment equation: additional things you can do to extract the maximum from your equipment and circumstances. Will everything in this chapter save you time? The answer is both yes and no. Because much of your viewing will probably take place in short sessions, failing to attend to a number of simple things will make it likely that you will spend much of your limited time grappling with avoidable problems. The frustration of dealing with recalcitrant gear can be enough to discourage you from trying to take advantage of such brief observing excursions in the future.

Even having the best, most practical equipment and top-notch observing skills do not complete the picture. Every link in the chain should be optimized for its best performance, which includes capitalizing on viewing conditions. On those rarer times when you are able to take your equipment to a remote site, it is no less significant to be able to take full advantage of those opportunities, too, by ensuring that your telescope can deliver maximum effectiveness and use, along with minimum downtime, tinkering, and hassle!

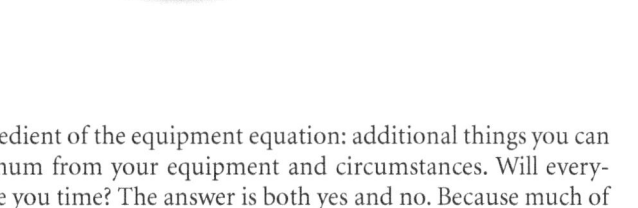

The Importance of Precise Optical Alignment

Attaining virtually perfect alignment of your telescope's optical components is something that cannot be stressed enough! Regardless of how much time you have at your disposal, this part of the equation is critical. You need to be sure that it has been properly taken care of from the outset so that your time at the eyepiece is not wasted.

A. Cooke, *Make Time for the Stars: Fitting Astronomy into Your Busy Life,*
DOI: 10.1007/978-0-387-89341-9_4, © Springer Science+Business Media, LLC 2009

Many newcomers fail to understand the degree of optical precision that has been crafted into their telescopes. Considering the microscopic tolerances of decent modern optics, the slightest alignment error introduced into the optical path will produce at least impaired, and at worst ruined, performance. To a newcomer, in the face of early disillusionment, the consequences may be an immediate souring toward astronomy.

The need to take care of the matter is no less important from one type of telescope to another; however, different designs and focal ratios make degrees of accuracy a greater or lesser issue. For the most part, for example, you need not usually be concerned with refractors; once set correctly at the factory, they rarely go out of alignment, unless subjected to a violent blow, or dismantling of the optical components by the overly curious. Correcting the problem is something you will not be able to take care of yourself, and it will require the services of a professional optical shop. Catadioptrics, such as the Schmidt-Cassegrain, are, more often than generally realized, prone to alignment problems; although they allow for easy collimation of their secondary mirrors, the same is not true of their primaries. Again, for primary adjustments, a professional optical shop will be required.

That old workhorse of amateur observers, the Newtonian reflector, while probably the most common choice among visual observers, is the type most likely to need more regular optical adjustments. This is true for virtually any of the mirror-based designs, although most of the variants, such as true Cassegrains and even the Gregorian design, are rarely seen these days. Fortunately, mirrors in these telescopes have benefited today from better mounting methods, so they tend to be considerably more tolerant to the things that caused constant misalignments in the past. Regardless, these designs allow ready and perfect collimation by their owners. Reflecting telescopes of low focal ratios are also more drastically affected by poor collimation; higher magnifications require ever-greater need for accuracy. Larger telescopes, especially of relatively short focal ratios, will perform like very poor instruments indeed when their optics are imprecisely aligned, even when the amount of poor adjustment is only slight. It will be hard to reach even a decent focus of planetary discs, let alone to resolve detail when the telescope is in such sorry condition! Therefore, do not neglect this factor, and spend as much time as necessary to reach the finest level of collimation possible. Time spent at this stage will reward you with huge dividends in results, and this is one area where you should never skimp, regardless of how busy you may be.

So, for the most part, Newtonians are the only type of telescope that we are able, or ever likely, to fully adjust ourselves. The only tool that has proven nearly indispensable in accurately collimating a Newtonian is a simple and inexpensive device known as a Cheshire eyepiece. Its use becomes increasingly important as the focal ratio of the telescope decreases, especially with those below F8. Perhaps the best solution is the combination sight tube *and* Cheshire eyepiece, which offers great simplicity of use. A good, inexpensive 1¼-in. (31.75 mm) one is obtainable from Orion of Santa Cruz, California. Meanwhile, I strongly advise against laser collimators, as the less-than-expert user may inadvertently introduce errors into the optical path when it appears that a perfect result has been achieved! Their use seldom produces any advantage, even when used by skilled operators, so they are best avoided, despite enthusiastic claims you may have seen or heard to the contrary.

Now, a collimation issue not typically addressed in advertisements involves sight tubes when applied to the collimation of larger, short focal length Newtonians. We have to consider that larger Newtonians have relatively large secondary mirrors. Their 2-in. (51 mm) focusers, usually low profile units, are selected in order to set them as close to the secondary as possible; otherwise, the secondary mirrors would have to be even larger. It seems that there are no 2-in. collimating tools commercially offered to the amateur, so when using a 1¼-in. adapter to accommodate these collimating tools it is normal to find that the end of their tubes eclipse some of the secondary's circumference, as well as that of the primary. Moving the collimating tool away from the secondary with extensions will help to some degree, but sooner or later it becomes obvious that the only way to address the situation fully would be to build a 2-in. collimation tool for oneself. However, it is easy to get exceptional collimation without resorting to such measures; happily, the standard 1¼-in. devices are just fine, once you understand their workings. There are just a few simple steps in using this tool; it may look complicated on the page, but there is really not much to it.

Quick Setup Project: Collimation Made Easy

Time Required: 15 min

There is way too much mystique associated with the collimation process in general, something that is really quite straightforward. Below is a complete guide to accurate setup, from the very beginning.

You can perform the initial placement of the secondary mirror with the primary mirror still covered, if multiple reflections confuse you. Move your eye close to the focuser. Rough center and place the secondary mirror so that it appears essentially in the correct place and orientation, and as a round reflection in the middle of the focuser tube. You should know that there are two possible configurations for the placement of the secondary mirror. If the secondary is offset, although it will appear circular, you will notice that it does not appear to be central to the primary, even though the illustrations in the instructions that come with the telescope will probably show it *right in the middle!* Instructions are frequently misleading about this highly crucial issue, which can cause considerable grief among the uninitiated. The reason correct secondary mirror placement should be this way is because one side of it is closer to the eyepiece than the other, meaning that the opposite sides of the cone of light will not be reflected equally.

The other possible position will result in the secondary appearing central in the tube. However, the secondary will appear slightly oval, nevertheless reflecting the primary mirror's center coincidentally to its own. More potential confusion! In centering the secondary with the focuser, if you measured it you would find that it is angled at other than 45°. No matter; the result and effect of collimation is the same. It is the final alignment that counts.

To adjust the secondary, install the collimation tool in the focuser, along with a 1¼-in. adapter if you have a 2-in. focuser. Uncover the primary. Ideally, the primary will have a spot, or better, a small central ring pasted centrally onto the surface. If no central spot is provided, glue a small paper ring on its surface at the center, which is a better option. Such a ring should have a central hole about the size of a dilated eye pupil. The center of the mirror can be ascertained by taping a fine thread to one side of the mirror, pulling it across the front surface to find its shortest point. Tape it to the opposite side. Take another thread, taping it on the side at approximately right angles to the first, and perform the same procedure. Where they intersect is the center of the mirror.

You may safely proceed even if you cannot see the edges of *either* mirror. You may have read that being able to see the edges of both mirrors is important, but as long as you work with the central mark this is actually what counts. Work any and all of the secondary's adjustments until the central spot on the primary coincides with the intersection of the cross hairs at the end of the cheshire accessory tube. Pay no attention to the reflections you see in the primary itself, and adjust the secondary only, in any direction, including turning it, as required to attain the end result. The secondary is now set correctly relative to the primary's axis. All other primary/secondary issues, such as the preset offset or angle of the secondary mirror, are automatically compensated for.

You may also notice a slight apparent slant of the secondary mirror's sides and holder in its reflection in the primary, seen as less-than-symmetrical around the secondary mirror itself, as well as similar effects seen in the sight tube/Cheshire eyepiece. This, in fact, is additional confirmation that correct placement of the secondary mirror has been achieved. Once accurate placement of the mirror is settled, try putting a small application of "super glue" on the secondary's central stalk, and on any of the spider's adjusting screws (if your telescope has these), as good insurance against any further movement. The seal is easily broken with the slightest turn of a screwdriver or wrench, so do not be concerned with future adjustment requirements.

Having set the secondary mirror, it is now a simple matter to adjust the primary mirror so that *all* mirror reflections center themselves in the cross hairs. The resulting reflections should appear like concentric circles, divided into the four parts formed by the cross hairs. If your secondary "spider" mount has four arms, you can even line up the cross hairs to fall in the same position as the arms. In order to simplify the entire second step, work with only two out of the three collimation screws of the primary, and leave one screw constant; this will eliminate much confusion and wasted time. Very small increments make huge differences with short focal ratios; an eighth of a turn is quite a major adjustment.

That, in a nutshell, is all there is to collimating a standard Newtonian.

You may recall the description of the solid construction of my own JMI reflector, in previous volumes, which makes recollimation a rare event indeed, even after I have taken it apart and reassembled it at a far away site! This is quite a validation of the quality and design of its manufacturing. In further testament to the great accuracy of its engineering, the stout spider arms are actually welded to the inside of the rotating nose portion of the main tube, eliminating the need for any further adjustment of the preset offset. Far from being a handicap, such adjustment will never be needed, although the secondary itself may be turned and raised or lowered relative to the spider.

In telescopes with rotating noses, initial setting of the secondary is an extra-complicated procedure. Any adjustments that you need to do are time consuming, because turning the nose to different positions around the circumference will initially reveal small differences in relative orientation of the primary and secondary mirrors, depending on any inaccuracies remaining in the secondary's adjustments. This is caused by very tiny divergences from what would ideally be a perfectly round telescope tube. Therefore, what may appear correct in one position will only be found to be slightly out of adjustment in another. It should be possible to achieve remarkably accurate placement, nevertheless. Averaging everything out can make for a very trying exercise in the initial setup, especially since the position of the secondary may have already been offset on permanent arms, without the possibility of further lateral adjustment. With any luck, everything will have been taken care of by the manufacturer and you will have nothing to worry about.

The same complication will occur in the collimation of the entire system as well, for the same reason. However, assuming accurate construction, rest assured that persistent tinkering should eventually achieve good overall collimation in all positions of the rotating nose, by gradually adjusting the primary mirror around the circumference until the differences are very slight indeed in all positions. Although I cannot claim that the final collimation will not be without the minutest inaccuracies, they should be so tiny that they have no effect on performance. In creating accuracy and stability, there is also the thickness of the spider supports to consider. Ideally built so that one may readily dismantle and move the telescope, hopefully they will be rigid enough that there is still no chance of introducing optical error! Relatively wide supports should be OK, as long as they are no more than a couple of millimeters. This will result in only slight extra light diffusion and diffraction; all told, a pretty good trade-off.

The Importance of Clean Optics

The ill effects of diffraction are not only the result of less efficient optical design but are also the result of dirty optics. Beyond a certain amount of normal dust inevitably collecting in only a short time on our telescopes' optical components, for the best results in viewing deep space we need to capture the maximum amount of light and also ensure that every possible photon of is transmitted and focused into our eyes. This means that we should try to prevent the telescope from absorbing or scattering any more of the light than necessary. So do not overlook the potential you might be losing with dirty optics, which scatter and absorb light all over the image plane and reduce contrast! Contrast is at no less a premium in deep space than it is in the solar system. Because of the faintness of most deep space subjects, simple maintenance of the maximum optical capacity of your telescope plays big dividends. So keep your telescope and accessories tightly sealed when not in use, *and do not be afraid of cleaning the optics periodically.*

For those who fear washing their telescope's optical components, rest assured that it is not scary or particularly difficult! More care is required than real skill, and certainly it should never reach a level of dread that detracts from the job. There are numerous methods of carrying out this procedure in the most uncomplicated way.

Regardless of whatever method you utilize, the single most important thing to bear in mind is, once the optical component is clean and still wet, *keep all of the standing water on the optical surface together in one pool* as you pour the water off the surface.

The simplest mirror cleaning process of all? After trying many approaches, all of which share similarities, perhaps the best and easiest method is described below. You might try it.

Quick Setup Project: Easy Cleaning of Optical Components

Time Required: 15 min

1. Wet down the optical surface when it is horizontally positioned. In the case of larger mirrors, do this on the floor of a standing bathroom shower. Then swill it with a healthy batch of warm water mixed with detergent. Gently work cotton swabs across its surface, in rotating spiral motion from the inside to the outside edge (change the swab a few times to keep any gathered grit from marring the surface), with no pressure of any kind, just the weight of the swab. Then rinse it off again with clean warm water from the shower. Follow this by gently pouring a clean jug of water over the surface; this will ensure that the water stays together as a sheet and be less likely to break up into isolated droplets.

2. Work diligently to prevent any droplets splashing onto parts of the surface that are already cleared of water. Gradually tilt the mirror at an ever-increasing angle while taking care that whatever liquid remains on the surface stays formed as a single pool. Then carefully pour off the water, while trying to keep it together in one body at all times. If you fail to keep most of the water together, repeat the process. However, if doing this is hard, the chances are that the optical surface is still dirty, with more gentle cleaning being needed. You should do whatever you have to do to keep the water from separating into multiple segments and droplets, which will only add to the problems of obtaining a completely unblemished finish. Once successfully completed, there should be virtually no water droplets at all left on the surface, and at worst only the tiniest ones, easily picked up with the corner of a paper towel.

All told, this should be no more than a 5-min job. Remember, it does not have to be absolutely perfect, even though it may well be! Tiny water spot imperfections will not have any deleterious effect on the performance of your telescope that you can detect.

Getting More from Your Newtonian

Even in this day and age you are still likely to hear the optical performance of Newtonians all too frequent berated. It would seem to come more from those who may never have used a fine, well-adjusted example, or those eager to promote more fancy (and costly) telescope designs. It certainly does not emanate from those of us who understand and actually use these venerable telescopes!

It is no coincidence that typically most top amateur planetary observers have used larger Newtonians for their best work. Any negative reputation still remaining emanates from the time when most amateurs built their own Newtonians, and fine primary mirror parabolas frequently existed more in their builders' ambitions than in reality! The mountings of these instruments often left a lot to be desired as well. Having seen quite a few examples of these telescopes, it is no wonder the telescope's design has often been blamed rather than its optical and mechanical quality, or adjustment.

So, along with poor quality, there are some other factors (along with suggested remedies) sometimes cited for Newtonians being less suited than refractors to planetary subjects and other high-contrast uses.

1. *Overly short focal ratios, resulting in increasing sizes of the secondary mirror*: This reservation is not without some justification. It may not be possible to alter the dimensions of the secondary mirror (probably already set at an overall optimum size for the specific configuration of your telescope), but you should at least try to avoid telescopes in which it is too large. About 20% or less than that of the primary is a good rule of thumb, most people being hard pressed to tell the difference in performance between this and with secondaries significantly smaller. The differences, while present, are not really visually noticeable, and it would be fair to say that a really fine short focal ratio Newtonian will equal a fine refractor at least 75% as large, or better. Such differences as there are will be more pronounced at higher powers, where the short focal ratio reflector may struggle to compete with the longer focal ratio telescope or apochromatic refractor. In a telescope with a secondary mirror of substantially less than 15% of the primary, most people will not be able to tell any difference at all between this and even the best refractors of similar apertures! However, one may not freely reduce the secondary mirror's dimensions, which, in a properly matched system, is chosen to intercept the entire light cone, plus a small margin to allow for the old and infamous optical defect, the turned down edge. On the other end of the spectrum, a secondary significantly more than 25% will be a problem and will result in a performance comparable only with much smaller apertures. Remember, this is one of the biggest drawbacks with catadioptrics, whose secondary mirrors often exceed 35%!

2. *Tube currents*: Although these are a well-known problem in closed tube Newtonians, you can drastically decrease bad tube currents with one or more well-dampened muffin fans blowing *across* the face of the primary and through open spaces on the opposite side of the tube. Even placing such fans behind the primary will be of some benefit – in short, anything that prevents air from pooling around and circling within the tube, and thus in the delicate optical path. Tube currents are probably the single most degrading factor in otherwise fine Newtonians. Solid metal tubes are among the worst offenders. Cutting vent holes all around the primary can help; even greater benefit will be apparent with muffin fans installed.

3. *Inadequate light baffling*: Poor light baffling reduces contrast. This may be addressed by using a light shroud around the large open spaces of latticework tubes, thin enough to allow for ventilation. The region in front of the eyepiece needs to extend far enough so that no light can enter the eyepiece tube from the

front end, too. A simple telescope tube extension (or shield opposite the eyepiece) can remedy this. If light is able to enter from behind the primary around its circumference, even a simple dark, aerated cloth can be used to block it out.

4. *Poor insulation*: Make sure that your telescope is insulated from warm air currents coming from you, the observer. With open tube designs, the same light shroud will also form a partial thermal barrier between you and the column of air inside the telescope. Good insulated clothing will help the situation. Insulation is a vital part of the equation, and far more a factor on image degradation than you may realize.

5. *Inadequate primary mirror support*: Thicker mirrors justify themselves in all circumstances and are wonderful performers, despite the added burden of weight. Modern thin mirrors, typically utilized in less costly Dobsonians, have the advantage in faster cooling times and easier portability. However, problems of thin mirror flexure generally remain significant in almost all mirror mount designs, even with mirror mounts featuring elaborate support systems. The task of figuring these mirrors during their production must create its own special problems for the same reasons. For viewing the planets, beware! Although we do not always require high magnifications in deep space, for the solar system relatively high powers are more the rule than the exception.

6. *Incomplete cooling*: Be sure to allow sufficient time for your Newtonian to take on the temperature of the outside air as completely as conditions allow. Naturally, the telescope will be most adversely affected during times of ongoing temperature drop after nightfall. While likely to be most severe in Newtonians, even the most "stable" telescopes types (refractors, and catadioptrics to some degree) need some time to adjust to temperature changes, so this factor is not all biased against Newtonians.

7. *Coma*: Far less problematic for live viewing than the detractors would have you believe, the negative effects of increasing coma toward to edge of the field of view are unlikely to bother most observers, who are probably more preoccupied with the central regions of the field than the edges. In any event, these ill effects may be negated by a simple coma corrector, if so desired, although you may never feel the need for one. In visual astronomy it is not very disturbing, especially since certain modern eyepiece designs reduce or eliminate it. Some eyepieces by TeleVue are well known for their coma correcting properties. Otherwise, the only people likely to complain about this optical trait are those involved in deep sky imaging. For them, perhaps the exceedingly good, though extremely costly, true Ritchey-Chrétien design is the best telescope design of all, being designed primarily as an imaging and not a visual telescope. It is certainly not the type of telescope for those short of time and opportunity to peer into the night sky.

All things being equal, there is no reason why the still sometimes maligned Newtonian reflector should not live up to its full potential, inch for inch. It remains for many the telescope of choice overall, especially in ease of use. Its clean image delivery takes a lot of beating by any other type of telescope. You could easily spend a lot more money for another type of telescope, which may be slightly superior in delivering the maximum for any given aperture, but likely will be far smaller; the result will usually be a lot less viewing potential, regardless. When it comes to what counts – excellent overall performance – if you have already spent any time with a well-designed and well-built Newtonian, these things speak for themselves.

For your own purposes, you will have to weigh the benefits and drawbacks of any telescope you are considering. If the planets are your main interest, high contrast should always be at the forefront of any telescope you are considering, which, for strictly visual applications, probably rules out most catadioptrics.

Other Distractions

In order to make your viewing as productive as possible, a particular "bugaboo" of many concerns the comfort level of using whatever equipment you have. Although it is not actually *necessary* to remedy any of these things in order to do worthwhile astronomy, these issues contribute more than their fair share of frustration to the average user. At the very least, time spent under irritating circumstances will mean less productive viewing, and you may gain nothing more than a desire to "pack it in" for the night. Worse, you may conclude that astronomy is too much trouble for your life and schedule, and that will be the end of it. Indeed, this particular problem is why many busy people cannot "find the time" to enjoy their chosen hobby.

In order to optimize your telescope to take advantage of any opportunity to use it, here are some things you should try avoid, or at least, to counter:

1. *Inadequate mountings and bases*: These transmit every quiver and twitch into the field of view; just when you try to focus something, it wiggles all over the place, even perhaps exiting the field! Then along comes a gentle breeze, and the image in the field of view will not stay still long enough for you to discern anything. Most novices fail to fully grasp the degree of motion amplification that even low magnifications impose on a poorly conceived mounting, no matter how well they are cautioned. Normal engineering tolerances are suddenly woefully insufficient; remember, a mounting should not only be mechanically strong, but in an ideal world, would be inflexible.

   Also consider the surface beneath your feet an extension of the mounting itself. If it is a deck with air space below, even a massively built one, the chances are that every movement (even just shifting your weight from foot to foot and without even taking a step) will be found ruinous to your viewing. (You can remedy such a situation by mounting the telescope on a concrete-filled pier, sunk into the ground below, which penetrates the deck but does not touch it.) Ultimately, there is nothing quite like a flat concrete base for your telescope's footing, although thick bases can suffer from lagging cooling times relative to those of dropping nighttime air temperatures. You might experiment with ordinary terra firma, but remember that there can be no allowance for any form of movement, however slight. A backyard lawn can be very good indeed but may prove problematic if the dirt underneath is soft or wet, or if the mounting's feet dig in ever deeper as your viewing session proceeds. There are good commercial footings available for setups of telescope tripods on grass; however, there is little to beat large concrete paving squares. Their effect is hard to duplicate, and you would be well advised to always take them out with you to the wilderness for dark sky sessions, where the footings are likely to be less than ideal.

2. *Distracting and vision-impairing lights:* These can ruin any stargazing session. There is a difference between general area light pollution and nearby bright lights. Both can ruin your viewing, but we have no power over light pollution. Image intensifiers react very badly to strong wayward lights. Maybe you can alter nearby lighting, though. To counter wayward nearby lights, you might try hanging baffles (dark sheets, or the like) from suspended wires. Some of your neighbors may be only too willing to help out, too, by turning off their own outside lights when you want to observe. However, there are also those who may not be readily open to the idea! Most people have absolutely no idea what our civilization has done to the night sky, and their own role in it; worse, they are completely unaware of this critical factor to telescope users, and even unsympathetic to our pleas. It is here that practiced skills as a diplomat may come in more handy than observing skills!

   For conventional viewing you will need to maximize whatever degree of dark adaptation is attainable; again, this may mean informing the uninitiated of observing protocol! Newcomers will soon find out why dedicated observers carry around red light flashlights. However, image intensifiers actually severely limit the degree of dark adaptation that you will attain (but you will not need it either!), so you should probably try not to mix one type of viewing session with another.

3. *Wind:* This creates its own set of problems and will not only ruin atmospheric stability but also shake even ruggedly mounted telescopes enough to become significant. It is less of an issue for deep space viewing (because of the lower magnifications and less critical resolutions generally needed), but it always makes for potentially unpleasant circumstances, particularly when chilly! (Memories of some wildly windy nights out in the desert come to mind, along with flying dust galore.) Regardless, image stability, if not the transparency of the sky, is degraded, but you can still carry out worthwhile deep space viewing. Again, aside from having the most rugged mounting possible, you might be able to take advantage of setting up lightweight baffles, perhaps panels of cloth – a temporary boxed in "observatory" out in the wild.

4. *Serious cold:* This is another thing entirely; unprepared, there is nothing quite like it to squelch (or should I say, even to freeze?) your enthusiasm. Sometimes things become so bad that any condensate forming on the optics in your telescope may become an ice sheet! When something truly exciting is in your field of view you are more likely to put up with discomfort, but nevertheless, your best times under the stars will probably be when you are physically most comfortable. Always have more than enough warm clothing, and do whatever you can to make your circumstances as pleasant and the least distracting as they can be.

5. *Sitting versus standing:* And what about the old argument for such comforts? Any uncomfortable contortion required to simply look through a telescope is a no-no, and it depends to a large degree on the telescope itself. Sitting is a nice relaxing proposition to be sure, whenever it is practical, but becomes a far less significant option when you are not forced to stoop, stretch, stand on tiptoe, or lean in some unnatural way. Most of the time you will find that adjusting a chair apparatus to the proper viewing position will be too much continual hassle, when a telescope better suited to your needs would have been the better choice. Because telescope design plays an important determining role in all of

this, be sure to research carefully what you are getting into with any particular instrument. Again, there is little to beat a moderate- to large-sized Newtonian for maximum comfort, especially those that place the eyepiece at standing eye level. It is unpleasant looking upward into anything.

The Weather!

There is one other factor in all of this that is also critical, and that is experiencing good conditions in the atmosphere itself. When time is short, who wants to go to all the trouble of setting up when viewing turns out to be unsatisfactory? However, aside from simply having a clear sky, your needs will be quite different from one type of observation to another. For the major "inhabitants" of the solar system, you must have steady air, above all. It does not really matter if the sky is light polluted or even thick with particular impurities or moisture. All that matters is that it is still. You will conclude therefore that the number of such optimal viewing nights is actually quite limited if you expect the best results. Rare is the time when all we see in the field of view is a slow "flip and flop" of the image here and there. It is made even worse as telescope apertures increase, as the width of the light beam entering the telescope tube is affected by the very width of the wavelengths of atmospheric motion. So there is, after all, a certain advantage to smaller apertures, because they will perform optimally more of the time than will a larger one! However, smaller telescope simply cannot outperform larger ones in the same conditions. Do not overlook this point, despite what you will hear some people say (usually owners of small aperture telescopes!). It is just that the larger one's advantage is increasingly and dispropor-tionately diminished as the air boils and stirs ever more strongly. However, one distinct advantage in the solar system is that we can do most of our best observing from home, no matter where that is. Needless to say, always use whatever resources are at your disposal to check beforehand on the viewing conditions you can expect.

For deep space viewing what you really need is dry, transparent air, as dark as possible, too. There is simply no substitute for either of these things, since they are the key to light transmission; this is the name of the game. While wind may not be especially damaging much of the time, when a need arises for higher magnifications air turbulence does become more problematic. Fortunately, much of our deep space viewing will require only low to moderate magnifications. So, although you cannot order up the conditions you need, you must do your homework really carefully before setting out on any grand venture to a remote site. There is nothing more frustrating, having lugged everything you have to a far away place, than to have to pack it up all over again and head for home: extra bad and discouraging if your time is limited! You can avoid most of these disasters with just a little care. So check all the weather websites you know, and never just hope for the best!

There is another ingredient to the mix as well. Because there is no way to know for sure what the weather may bring, always be prepared for the worst; a sudden unexpected rainstorm can not only ruin your viewing but destroy your valuable equipment as well. Thunder and lightning, or even hail, may be even worse! You may not have time to dismantle everything and move it to safety before disaster strikes.

Even if you do have time, you will then have to go through the setup all over again. So, while you probably have some type of protective covering for your telescope, make sure it is truly waterproof, and that it will fully cover the equipment and will not blow away easily.

In the USA, Mexico, and Canada an invaluable astronomer's resource of hour-by-hour weather predictions for most locations, with particular reference to conditions critical to viewing, can be found at http://www.cleardarksky.com.

The Value of True Portability

Here are a few more comments about the importance of your equipment having real portability:

Figure 4.1 shows my own JMI 18-in. (45 cm) telescope fully set up at a dark sky site in the barren but awe-inspiring California high desert. Although the complete process, start to finish (from loading at home to full setup at dark sky site) is taxing to be

Fig. 4.1. An 18-in. scope, set up in the desert.

sure, all minimum prerequisites are preserved in this particular telescope. Assuming that you do not want a telescopic white elephant that will only end up in storage, you might want to seriously consider such similar good design traits in your own equipment choices. You will find various designs by different manufacturers that incorporate most of them, albeit differently in the details, so such telescopes should not be hard to find, even though not usually seen in commercial showrooms.

If you elect to have a telescope with full equatorial capability, how much precision does the equatorial setup need to have for your particular type of use? Fortunately, for most live viewing, you will not need to achieve more than a reasonably close polar alignment; just lining up the pole star fairly closely with the telescope tube is more than sufficient, and the tracking is surprisingly effective. Only for long exposures does the need for perfect tracking become a necessity, where the slightest wandering of the image over even a fairly long period is critical.

In the past, such large amateur telescopes as we see today would have been strictly fixed observatory models, weighing much more would be practical for portability. Amazingly, despite the slightly slower damping rates of these new instruments from any vibration, the better examples will have few disadvantages over much heavier ones. At a remote site, it is desirable for you to be able to complete the unloading, and complete setup in not much more than half an hour. It is quite normal to find the optics of good modern telescopes in perfect collimation after total disassembly, travel, and reassembly! It would seem that some things, in the hands of the right manufacturers, have again certainly changed for the better, and they are all important benefits for the type of astronomy we wish to participate in.

However, despite the attributes of many larger modern designs, having two people rather than just one to move them to remote sites makes things a whole lot easier, so there is a finite upper limit for the single observer. Most solo observers would probably prefer to keep their telescopes to a maximum of 12-in. (30 cm) reflector aperture or so, simply because of the difficulties before the setup even begins! Refractors would need to be much smaller still, as comparable apertures would definitely not fall within the bounds of portability.

It is possible to have great performance attributes for a fraction of the cost and size of the grander telescopes. Optical quality may be the most difficult ingredient on which to skimp. However, many lightweight mirrors are available today, and although they will be unlikely to equal the quality of their more massive cousins, they can be more than serviceable if properly mounted and supported. A good deal of cost may be shaved off the price by utilizing only *somewhat* smaller apertures, since costs increase diametrically with size. The ratios of aperture versus bulk are also dramatically affected with each decreasing inch of aperture. To save further, basic Dobsonian mountings are simple, smooth, and stable if you can live without ready equatorial tracking.

Section II
The Moon

CHAPTER FIVE

The First Port of Call

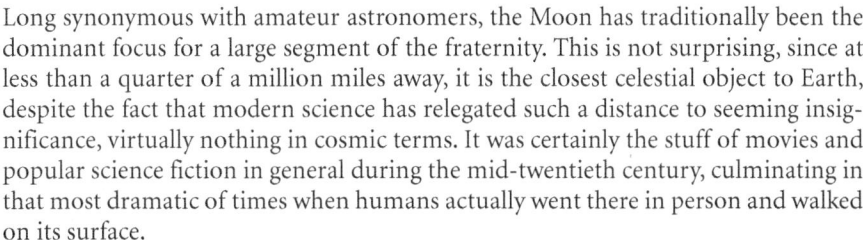

Long synonymous with amateur astronomers, the Moon has traditionally been the dominant focus for a large segment of the fraternity. This is not surprising, since at less than a quarter of a million miles away, it is the closest celestial object to Earth, despite the fact that modern science has relegated such a distance to seeming insignificance, virtually nothing in cosmic terms. It was certainly the stuff of movies and popular science fiction in general during the mid-twentieth century, culminating in that most dramatic of times when humans actually went there in person and walked on its surface.

After finally having provided such an astonishing realization of a long-held dream, all too soon the Moon found itself banished to the attics of popular interest! With nothing left to excite the imagination, the very success of the *Apollo* missions is largely why most amateurs set aside their traditional interest in the Moon. The sensational imagery from these missions, as well as those of all the orbiting spacecraft, made our old friend seem humdrum and too familiar for the amateur observer. From the beginning of the Space Age, the Moon has been mapped and analyzed in astounding detail (and from all angles).

Nevertheless, the splendor, the ready viewing potential, and the almost unbelievable detail we can see in the eyepiece remains unchanged. It is all too easy to forget, or overlook, the fact that live lunar viewing provides a sight and opportunity that nothing else can begin to duplicate! And through the telescope the Moon still is the only place outside Earth's environment where we may actually feel as if we can touch another world, and on almost the same terms with which we experience our own.

A. Cooke, *Make Time for the Stars: Fitting Astronomy into Your Busy Life,*
DOI: 10.1007/978-0-387-89341-9_5, © Springer Science+Business Media, LLC 2009

Just as in viewing the Sun, good lunar observations do not necessarily require large telescopes. The apparent size and brightness of our satellite makes up for many a shortcoming in equipment. Compared with the ever-larger equipment that we will need to see the rest of the universe with any reasonable degree of success, even a child's telescope will give a novice a thrill when pointed at the Moon. Best of all is the fact that we can see a great deal in very short observing sessions. With fairly good viewing conditions, and even just a few minutes or so, we can be transported to another destination, so convincingly that we might as well be there! There is no straining to see a wealth of detail in the field of view, no waiting for perfect conditions to have any chance of enjoying something worthwhile, and certainly no requirement that we have the most expensive instruments to see it to advantage. So do not overlook this greatest of all natural viewing opportunities.

For countless people, the Moon was certainly their first astronomical destination, and probably enough to set many of them into a lifelong hobby. It is hard to believe nowadays that this little world held so much of the collective attention of so many observers over the ages, and popularization of the telescope, then at last spaceflight itself, allowed all of us to share in it.

In one of the most enthralling of all telescopic views, looking at lunar limbs near the terminator can simulate flying by the Moon as if in a spaceship. If we want to experience our own natural satellite with something akin to such a "manned spaceship" vision, we will need optical excellence of at least a decent, but not necessarily great, aperture, although larger telescopes of high quality give us virtually unlimited potential and increasing realism. Thus, with the well-illuminated views that even average telescopes can provide at higher powers, we can indeed realize something of a "lunar fly-by," and feel we are actually close to the surface.

Quick Project: Lunar Fly-By

Time Required: A Few Minutes!

It is best to wait for a night of at least reasonably steady viewing; a boiling atmosphere will destroy very quickly any illusion of being in space! Try to line up a viewing angle that resembles the vantage point of being in orbit, where you are looking nearly parallel to the surface toward the lunar limb. When viewing the Moon at such an oblique "approach," the sensation of flying by it is actually quite physical. Add to it the dramatic, almost three-dimensional appearance that the Moon takes on in the field of view. The effect is even more striking with binocular viewers, and noticeable even with relatively small apertures in a way unique to the Moon. Using electric drive controls to pan across the surface only further adds to the illusion.

Although viewing anywhere near the terminator will suffice, for the best results of all you need to look toward the limbs, just before or shortly after full Moon. It is also possible to appreciate similar perspectives at the poles, as long as a sufficient portion of the disc is illuminated. However, the effect is usually less dramatic because the

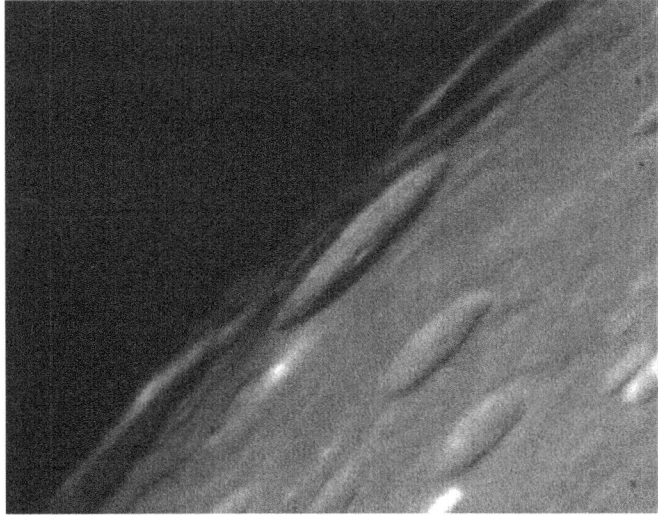

Fig. 5.1. "Fly-by" over La Perouse (image by the author: JMI 18-in. telescope, Astrovid 2000 CCD video camera).

lighting is not so favorably placed, being at right angles to the former. Features near the lunar limb evoke a sense of actually being in close orbit better than anything else, because we are positioned as if looking toward the lunar horizon from a spacecraft window. It is at these times that the gentle undulations of lunar mountains may also be effectively observed.

With such telescopic opportunities as we have, a beautiful and refined sight, the crater La Perouse (10.7°S, 76.3°E) certainly provides a perfect subject for the illusion (Fig. 5.1). Looking very much like a perfectly formed saucer with a central bump, it was imaged just after the full Moon, on the eastern limb and quite near the darkness of the encroaching terminator itself. It really does feel as if we are there! In many ways, we are. See also Fig. 6.1 (Chap. 6) for another example.

A Real Lunar Fly-By!

Certainly, there are countless photographs from the *Apollo* missions that show incomparably detailed orbital perspectives as seen by astronauts from lunar orbit (Fig. 5.2). Here, just a handful of people through all human history experienced true lunar fly-bys, actually living many an old dream!.

Prinz is a fine example of a crater that has been flooded by lava flows of ancient times, filling in its inner walls almost to the point of total obliteration. Notice the almost countless tiny craterlets contained within its ruined walls, in addition to all

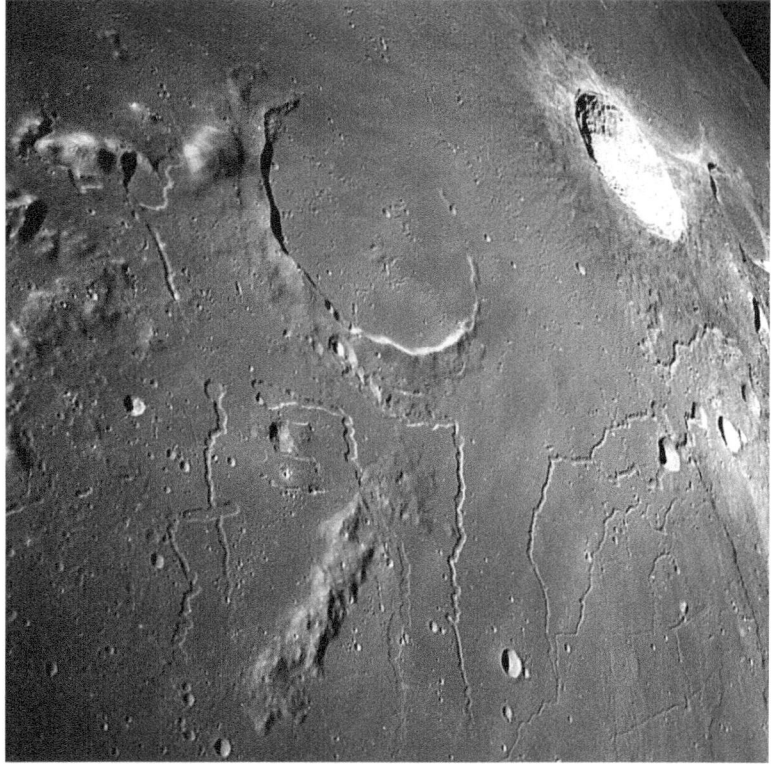

Fig. 5.2. Orbital view of ruined crater Prinz (27°N, 43°W: *Apollo 15* mission); the well-defined crater nearer the horizon is Aristarchus (photocourtesy of NASA, scan by Kipp Teague).

over the surrounding plain. Also note the clearly flat-floored rilles, a product of collapsing lava tubes, rather than running surface liquids.

The magnificence of the true lunar "fly-by" in the aforementioned illustration is most impressive here. Of course, we have far fewer opportunities to create these orbital illusions from Earth, because we can only take advantage of features on the extreme limbs; actual orbiting spacecraft have continually renewing horizons! However, in our own more humble efforts we can certainly experience some of the same wonder and awe the astronauts must have felt.

But now it is time to look further than our nearest neighbor.

CHAPTER SIX

The Moon: Close Up and Personal

Because so much excellent material about the Moon is widely available, this book does not intend to parade a restatement of now familiar themes and information, already superbly laid out by some prolific lunar authorities. Of the many printed resources available for the amateur observer, there is the excellent book *Observing the Moon* (Gerald North: Cambridge University Press, 2000) an outstanding reference volume by a tried and true lunar observer. It is certainly one of the best, if not indeed the best of its kind for the amateur observer, containing in its numerous pages highly illuminating, imaginative, and detailed methods to view and study the Moon's surface.

You should also have a good lunar atlas. Surprisingly, there are few really first-rate choices for the amateur observer available, although *Hatfield's Lunar Atlas* (a veritable classic dating from the 1960s and now reprinted by Springer) probably comes closer to the mark than most, despite its old, somewhat obsolete photographic imagery. Times have changed, and advanced technology has certainly given us vastly improved methods of imaging the Moon. Too bad a lunar observing authority has not yet compiled and printed a new atlas along Hatfield's lines. There is also the classic *Atlas of the Moon* (Antonin Rukl: Kalmbach Books); while a masterpiece of cartography to be sure, there is something about these incredibly detailed drawings that does not translate ideally to our perceptions when we are scouring the lunar surface. Add to this the strangely clumsy and problematic layout, which makes referencing during telescopic navigation of the Moon's surface less than straightforward. With time always pressing, we need easier lunar navigation!

Maybe another good book in which to invest would be *Full Moon* (Michael Light: Knopf, 1999), a wonderful photographic compendium of lunar sights as seen by the *Apollo* astronauts. From the perspective of comparing such views with what we see through the telescope, the selection of imagery from the *Apollo* missions,

A. Cooke, *Make Time for the Stars: Fitting Astronomy into Your Busy Life,*
DOI: 10.1007/978-0-387-89341-9_6, © Springer Science+Business Media, LLC 2009

again, is highly revealing. This is because in this grand coffee table size volume, so many of the book's magnificent illustrations were taken on or near the lunar surface. Although its images are not nearly so extensive as those available at the NASA web site, it does provide a quick reference and beautiful crisp imagery, presented in a way that only the pages of a book can. Through such imagery, we are able to better appreciate the surface of the Moon, often photographed directly above from lunar orbit, instead of only from the unchanging angles of our own telescopic views.

The Moon is still the *only* destination in space that people have actually photographed "on location." This alone provides reason to return again and again, via our own private "spacecraft" (our telescopes!), because we can gain an even fuller appreciation, armed with new insights from spaceflight. With this in mind, the Moon may actually look even better to you now, and perhaps you may find yourself sometimes looking at it again with the same wonder that started you out. The Moon's frequency of appearance in the skies, ready access, together with the constant variety of detail, forever presented in different lighting conditions, make it a prime *and ready* viewing opportunity we should appreciate and not overlook. However, perhaps the greatest lure for us is the ready and virtually immediate access the Moon allows us, with none of the straining at the eyepiece so prevalent for seeing virtually all other subjects in space. If time is not on your side, the Moon is!

Choosing a Telescope

For any serious lunar study you will need a telescope of at least a reasonable size in order to pick up important details, although what you may see through even quite small telescopes might surprise you. However, a really fine telescope of, say, 6 in. (15 cm) or even less with exquisite optics will reveal our battered and cratered companion in stark and stunning detail. On a steady night, through larger telescopes such as a 12-in. (30 cm), or larger, say, 18-in. (45 cm) reflector, the views are indeed startling; the mind boggling complexity and refinement seem quite impossible to describe or image successfully. Thus, in order to see lunar sights in the eyepiece with the same clarity and realism as shown in the images here, you should realize that you will *not* need a telescope as large the 18-in. telescope that was used to illustrate this book. In fact you will probably see it better with far less! You will see the Moon far better in person than anything on these pages, because something is always lost in the imaging and printing processes, regardless of the type of system we employ. This remains true also of other imaging systems far more sophisticated than that used here. There are indeed some unique qualities in which the eye alone still reigns supreme.

The Moon is so generous with its light, however, that when viewing it through anything more than small apertures, and at low to medium powers, you are likely to find yourself being dazzled, literally dazzled; the impact seems all the worse once you take your eye away from the eyepiece. The intensity of the blast of light overwhelms the eye, leaving it impaired for many minutes afterward! Ever larger amateur apertures amplify the lunar glare even more noticeably, and at these times a lunar filter will be found to be more than beneficial (actually, a necessity!). Unfortunately, the resulting view is less pleasing than natural views. However, you will not need such a filter for high-power viewing. So, in relation to the Moon, with these larger telescopes it is sometimes better to wait for really steady air and go for really high powers, which the Moon will handle better than most subjects. (Of course, you already know of another remedy for glare: stereo binocular viewers, the use of which will restore much of your lunar viewing and provide a remarkable degree of comfort. Some observers also recommend various colored filters for viewing intricate detail on the lunar surface, but such filters remain a much overused option for live observing in general, the effects being decidedly unpleasant on the Moon and rarely helpful.)

In relative terms, it is always more difficult than it sounds to comprehend what we are looking at in context through the telescope. Using the most basic of comparisons, it is a problem to appreciate, say, a tiny ridge or crater when referencing only standard measurements for descriptions. Even the smallest visible lunar feature viewed at high power is still far from an insignificant size, compared with our own earthbound references of measurement. Far easier to digest is comparing a given feature to a well-known earthbound landmark or other feature of reference. By mentally super-imposing something familiar on a given lunar feature, this in turn gives a meaningful and personal insight to the whole observing experience. It will actually help you to "visit" the lunar surface itself. The appreciation of what you are witnessing will be changed forever by this extremely enjoyable pursuit of near limitless potential. It also will force you to realize just how big the supposedly "tiny" world of the Moon actually is, something we may routinely dismiss as being only a quarter the diameter of our own planet. The sizes of specific formations become even more significant when we consider that the Moon's entire surface is dry land. Because so much of Earth's surface is ocean, it means that the dimensions of lunar formations may be far larger than we suppose. Earth's total dry land is less than twice that of the full continuous dry surface of our neighbor.

Quick Project: Comparing Lunar Features to Familiar Landmarks

Time Required: 5 min

This is an easy one, but just as easily overlooked; for us, it is an ideal opportunity. Choose and compare a feature on the lunar surface to a familiar landmark. The fun of the exercise is to find something ever smaller in the field of view (for example, a craterlet on the floor of Plato), and reference it against something, such as a large city block. Such tiny features will no longer seem so tiny! Consider as well, perhaps, contrasting such monuments as the Eiffel Tower against the "Straight Wall" (about the same elevation), or a small but maybe significant landmass, such as Gibraltar or Manhattan compared with similarly sized lunar features. Even a crater, such as Clavius, which is far larger than most, can be put into perspective when you visualize its dimensions as approximately the same as the distance across greater Los Angeles, or the journey from London to Bristol.

Flying with *Apollo*

These days, when viewing the Moon, there is much pleasure to be found in revisiting some of the *Apollo* landing sites. This is something you can do without allotting large amounts of time, because the Moon is very observer "friendly," and most of the total resolvable detail may be seen almost immediately. As the years march on, it is becoming increasingly difficult to imagine that people actually walked these places. Spend a little time looking up at a full Moon one night, and this feat begins to seem even less likely. No wonder certain eccentric or uninformed individuals have periodically questioned that we ever went there, and what we are seeing are merely photographs of movie sets!

In the days prior to the Moon landing program, we could only speculate on the true nature of the lunar surface. Although we knew a lot about the Moon, our knowledge remained incomplete. Nowadays it is possible to gain a really accurate impression and perspective for ourselves, by taking advantage of all that is known in conjunction with our own viewing. By comparing your own observations with official NASA photographs and charts from the archives of lunar missions (see Chap. 16 in this book), you will gain an entirely new perspective on this old friend. Although visions of the "jagged cliffs" of the lunar surface may have long ago been shattered, it is said that the truth can set you free! In many ways it is more wonderful, because what we can see now is real, and not imagined. It finally is possible to have a real sense of how it would be to stand on the surface.

The photographs taken on the surface finally lay to rest any misconceptions of how the lunar surface really looks, which so often had been inaccurately portrayed in artists' conceptions of the past. In times not so very long ago (actually right at the dawn of the Space Age), seeing artistic impressions of how it was expected to appear was commonplace. It seemed that they were always featuring jagged peaks and ridges, all dramatically and sharply defined. The misconception existed even among some of the most informed people, almost until the day we touched down on its surface! How wrong all of this was, and how wise we always think we are, in hindsight!

The illusions of jaggedness on the Moon appear this way because of something akin to the Foucault test utilized in telescope mirror making, where lateral imperfections (measuring, at least we hope, no more than millionths of inches!) are thrown into exaggerated relief. On the Moon, the sharply defined, lengthy shadows thrown by the projecting or depressed features are also enhanced and highly contrasted without being diffused, thanks to the lack of atmosphere. The extreme lengths of the shadows are thrown on the surface by the low altitude of the unfiltered Sun, only just above the lunar horizon, rather than by any extreme shapes of the features themselves! In fact, most of these lunar features are more like rounded blips and undulations on a vast surface than the towering and sharply sculpted formations they appear to be from afar.

However, understanding this optical effect only partly explains why the popular misconceptions prevailed for so long. Even a casual study reveals, especially considering the enlightened state science had reached even 60 years ago, that those distorted ideas of dramatically tortured lunar features should never have come into popular acceptance in the first place! It is remarkable just how the perspective of the simple video image in Fig. 6.1, taken through an 18-in. telescope, corresponds to the scenes witnessed and recorded by the *Apollo* astronauts!

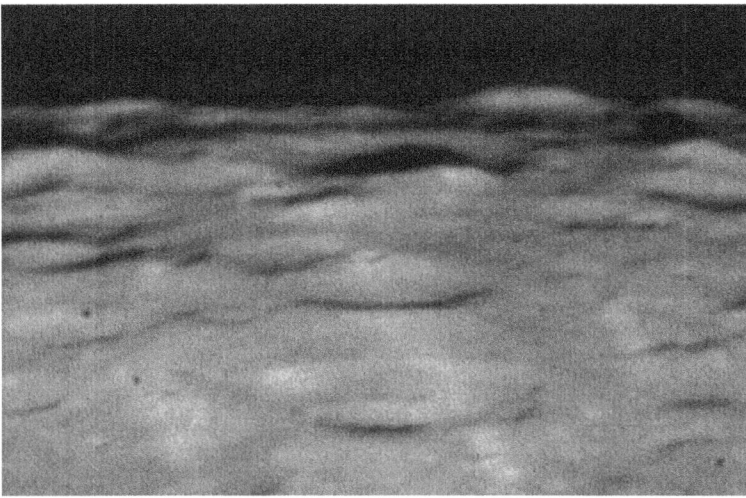

Fig. 6.1. Lunar mountains, undulations, and "rays" on the lunar limb (single CCD video frame:18-in. reflector).(AC)

Quick Project: Examining Mountainous Contours at the Lunar Limb

Time Required: 5 min

It is strange to think that it was always possible to have a pretty good idea of the true appearance of the lunar surface. We should have been well prepared for the type of landscape we would find at the surface! Just try it! Nothing could be easier or more demonstrative.

Look through your telescope at the mountainous detail revealed at the lunar limb just before or after the full Moon, since this corresponds closely to the horizontal view-point. Using a reasonably high power, you will find that the surface closely resembles the views from low lunar orbits of the *Apollo* missions! You will see no jagged edges, no sharply projecting points, no towering peaks – just endless undulations and smooth curves (see Fig. 6.1). You are now comprehending the true nature of the lunar surface.

Apollo

The celebrated *Apollo* landing sites provide more fodder to affect all of your prior perceptions of nature of the Moon's surface. Every landing site is worthy of this approach, and while the natural first inclination might be to check out the historic landing site of *Apollo 11*, it is relatively plain from a telescopic point of view. In fact, this first landing site was selected precisely because of such characteristics, in order to minimize risks from any unknown or unforeseen hazards on the surface. (Some scientists had even speculated that, like stepping into quicksand, the lunar module or astronauts would disappear into a thick layer of lunar dust!) This is not to deter you from viewing it, but you will probably find later *Apollo* landing sites to be even more interesting, since NASA's growing successes, confidence, and finesse enabled it to select ever more challenging locations.

When you visit the NASA web site and first begin perusing the image collection, you will find perhaps that the biggest immediate challenge is to just sift through all of the images brought back from these missions! However, these documented records contain all of the imagery of *Apollo* – quite a resource. They will certainly provide more than enough enlightenment and entertainment for any rainy night, and probably much more. The best part is that you can carry out your research at any pace; this is up to you. It can be quite confusing to try to make sense of long familiar lunar features, seen from entirely different perspectives, but compare telescopic viewing to looking down on our own planet from an aircraft. To think one would gain a true impression of any place on Earth from this perspective alone would be foolhardy, and such in-flight views above Earth's surface are certainly much closer to the surface than the apparent altitude of any telescopic or orbital view of the Moon.

Hopefully, with some new orientations firmly planted in your mind's eye, you will be more able to appreciate the lunar surface more fully, and similarly "put" yourself on the surface at the various other *Apollo* landing sites, or any destination on the lunar surface in general. Therefore, do not stop your adventures with the ready-made results shown in this writing; explore all of these *Apollo* sites for yourself by visiting the NASA website, but be sure to have your lunar atlas in hand. Spend however much time you choose, but at least visit the site!

Perhaps the most striking thing you will take from these NASA images is how nearly impossible it is to grasp any sense of scale or distance in them; a more telling testament to the lack of atmosphere and relatively pristine state of the lunar surface itself was never more clearly demonstrated! The pictures often do not look real. Aside from the gentle rounding effect on the lunar features, caused by billions of years of continual meteoric rain and cosmic debris, the surface has not been molded by such earthly phenomena as water, atmospheric winds, and rain storms, with all the consequential sculpting and eroding of landscapes that we take for granted in our own world. Plus, apparently there has been little seismic activity on the Moon since the earliest times. (However, recent studies at Brown University indicate at least a little recent and possibly limited ongoing volcanic activity.) In this latter respect, you will notice a relative straightforwardness of the lunar rock strata, as it occurs consistently layered and angled even through entire mountain masses. Compare this with the typical fractured and multiangled strata of mountainous formations on planet Earth, often stacked in many directions after a history of seismic turmoil. Although we can see that at one time the Moon was indeed a turbulent and heavily bombarded place, it is also equally clear how silent it seems today, and that it has been that way for a very long time.

Revisiting Familiar Lunar Features

It is quite possible for you to experience, quickly, and in an almost firsthand sense, what it must have been like to set foot on the Moon. With just a little patience, a little preparation, and certainly not a lot of time required at the telescope, the approach outlined later will undoubtedly change your attitude toward the Moon forever. You can relive these voyages repeatedly, whenever you want; they will never become humdrum. Such comparisons provide new insight, bringing what you are seeing to life and making your lunar visits infinitely more rewarding. Certainly they maximize the effectiveness of the time you have to spend and save you ambling aimlessly across the lunar terrain. We have all been guilty of this; how unsatisfying such forms of viewing quickly become! The best part is that you can accomplish a great deal quickly and effectively, even in conditions ill suited to viewing other destinations in space.

Quick Project: An *Apollo* Mission Relived

Time Required: 15 min of Observation per Landing Site (Plus Relevant Online Research)

This is just one example of the kind of discovery that you may experience easily for yourself at any landing site. Exactly how much time you spend will depend on your level of interest and, of course, the time available to you. Fortunately, it is quite easy to split your information gathering over many short sessions, so you can search first and view later! You could expect to have a pretty comprehensive grounding prior to viewing any site in not more than an hour total. As a process, briefly view the site first, taking as many mental notes as possible, then do your research as suggested, at a later date again returning to the same site with your telescope.

The landing site of *Apollo 15* is one of the more visually dramatic sites to contrast with earthbound observations. It is a wonderful and varied landscape and had long been a popular "destination" for many earthbound observers before *Apollo.* Thus, as an area of considerable topographical interest, the *Apollo 15* landing site lies nearby the long chain of mountains known as the Apennine range and is flanked by Mount Hadley (27°N, 5°E), Mount Hadley Delta, numerous interesting crater formations, and, of course, the now famous Hadley Rille. The site promised and delivered much to the visiting astronauts, who prepared a very complete record of the most well-known attributes of the area from ground level. In presenting a brief progression of photographic examples from the vast array of NASA's available *Apollo 15* imagery, you should quickly gain your bearings. Although there are many other photographs that might have fulfilled the same purpose, these show the kind of approach you might take as a basis for your own future explorations.

In the images later (Fig. 6.2a–c), the Lunar Module of *Apollo 15* was scheduled to land just beyond the three-sided turn of Hadley Rille (right center), on the flat terrain between the two groups of craters, labeled South and North Complexes (c). The hilly formation, labeled Hill 305 in (b), and seen at the upper right in (c) may also be seen on earthbound photographs as the northernmost point of a much lesser range running parallel to the Apennines across the wide and flat valley separating them. The smaller feature Bennett Hill, labeled in (b), may be seen easily in (a), adjacent to the crater Hadley C. Seen in the illustration (c) taken from lunar orbit before landing, the mass to the left is part of the Mount Hadley Delta formation.

You should by now have your bearings (the angle of approach may easily be compared with views from Earth). We now descend to a much lower altitude and to an entirely new perspective (Fig. 6.3). All of a sudden, the view is nothing like any

Fig. 6.2. (a) Hadley Rille and environs, familiar to most earthbound lunar observers. This finely resolved view is from high orbit (courtesy of NASA, scan by Kipp Teague, *Apollo 15*). Mount Hadley is prominent at mid-upper right. **(b)** From a closer range, here is a lunar satellite image with named landmarks (courtesy of NASA). Rotate 90° clockwise for comparable alignment. **(c)** Region seen obliquely from lunar orbit includes named small landmarks. This is the view facing northeast of *Apollo 15* landing site (between South and North Complexes at lower center). The three-sided turn of Hadley Rille is easily recognizable here, despite the different orientation (photocourtesy of Jim Irwin; NASA).

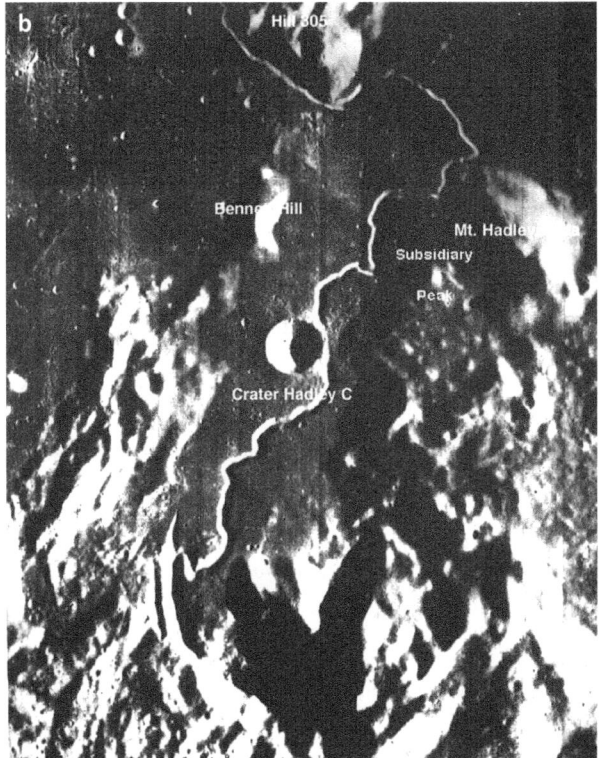

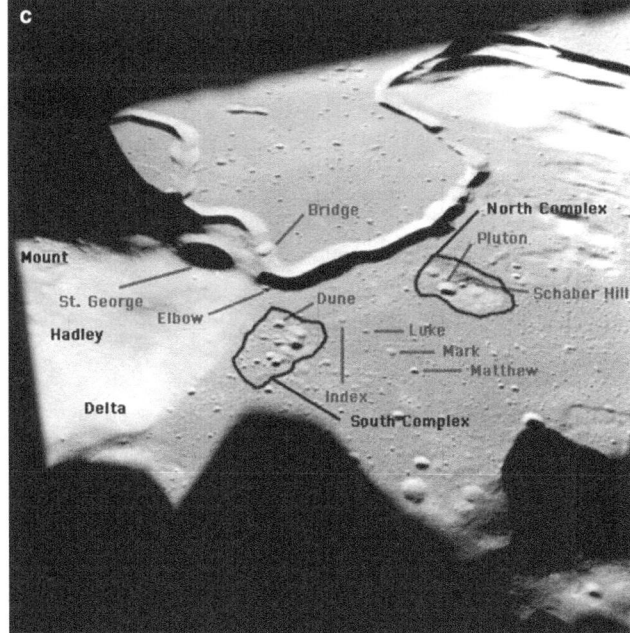

Fig. 6.2. (continued)

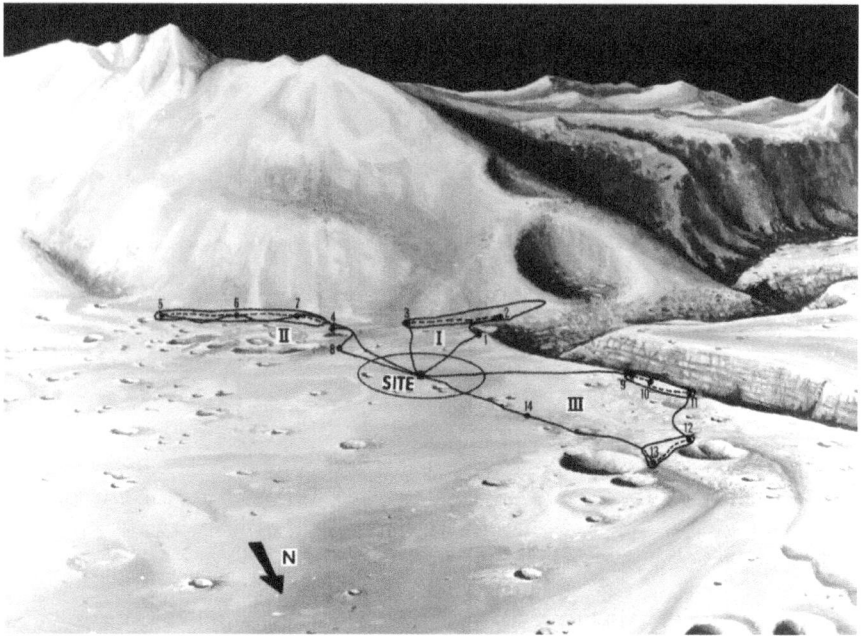

Fig. 6.3. (Photocourtesy of NASA, scan by David Harland). Landing Sight sketch projection.

that we will see from Earth. This preflight sketch projection of the *Apollo 15* landing site is more like a barren and rocky landscape in a remote desert. The proposed rover excursions are shown facing toward Mount Hadley Delta. They give a very good indication of what we might see from an elevation similar to the nearby mountain ranges. Many of the *Apollo 15* mission photographs were taken from these locations during the mission excursions. Note Hadley Rille making its sharp turn by St. George Crater on Mount Hadley Delta slope. Compare this with Fig. 6.2c, taken during decent from orbit a little more to the east. At this stage it should now be fairly easy to visualize where you are, compared with more familiar views. St. George Crater is a good reference point, as is the near square three-sided turn of Hadley Rille adjacent to it.

In Fig. 6.4, with the remaining descent over, we have finally landed on the Moon, here. Try to place yourself mentally somewhere near the region labeled "Site" in earlier Fig. 6.3. Thus, the view here (Fig. 6.4) is what you would expect to see on the surface: a wonderful panorama of Mount Hadley Delta and the striking "Silver Spur" feature (jutting out and up to the left). Note also St. George Crater on the lower mountain slope at the far right, and also what appears to be part of the group of craters labeled "South Complex" at left.

Turning 90° to the left, the Swann Hills formation (named for the *Apollo 15* geologist), along with the grand rock of Mount Hadley itself, looms into view (Fig. 6.5).

Fig. 6.4. Mount hadley delta mid 'Silver spur' (Photocourtesy of NASA. Image assembled by David Harland and scanned by Kipp Teague).

Fig. 6.5. Mount Hadley, left. Part of the Swann Hills range may be seen to the right. The rough-edged rock in the foreground of stands in contrast and splendid isolation, befitting the lonely stillness of the place (photocourtesy of NASA Johnson).

Although these appear to be quite small and relatively close in these illustrations, it is the lack of atmosphere and simple geologic strata that are playing tricks with our senses. The base of these formations is an estimated minimum of 4 miles away from this vantage point. Considering that their summits would compare favorably with many grand elevations on Earth, their structural simplicity make them appear less mountainous and more like small, gently rolling hills. However, this is the way of things on the Moon!

Turning our attentions now to the now famous Hadley Rille itself, something that had always appeared diminutive and unspectacular in our telescopes, we realize now that it is much larger than any prior impression we may have had. At about 1,000-ft deep by half a mile wide along most of its length, the term "rille" is not one likely

Fig. 6.6. Hadley Rille, looking north from base of Mount Hadley Delta near Elbow Crater. This location is close to the rille's sharp right angle turn (photo courtesy of NASA Johnson).

to have been used for such a feature on planet Earth! (Hadley "Canyon," perhaps?) It will immediately be apparent why the *Apollo* astronauts could not have simply crossed it to explore terrain on its other side. Again, it is hard to gain any sense of these dimensions from the stark illuminations on the Moon.

We now continue our sortie along the base of Mount Hadley Delta; it is extraordinary to think that this is the very same feature we see so plainly in our telescopes, and which appears so slight and unimposing. The same segment, shown so prominently in Fig. 6.6, with its protrusions back and forth across the rille, is still visible (at right) in the distance. Hill 305 (see b in Fig. 6.2) may now be seen in the background, to the left (Fig. 6.7).

Fig. 6.7. Hadley Rille from the base of Mount Hadley Delta (photocourtesy of NASA Johnson).

Fig. 6.8. Hadley Rille interior (photocourtesy of NASA).

Apparently Hadley Rille is typical of most lunar rilles and would seem to be a collapsed lava tube, rather than any form of river channel or such, because there is no evidence that water ever flowed on the Moon. Its appearance close up (Fig. 6.8) bears this out, since there is no sign of flowing erosion; rather, the rocky sides of the rille seem to have slumped straight down to form the channel. (Careful; do not lose your balance!) At around 1,000-ft deep, it is a lot further down to the bottom than it appears! Note the strong three-dimensional effect created by the out of focus near side rim of the rille along the lower portion of the photograph.

A near miraculous photograph (Fig. 6.9) sums up the mission as well as any other. Looking north we see not only the lunar module but also the crater Pluton (see again Fig. 6.2b) behind. What is easy to regard as a minor crater (which is what it is!) suddenly takes on much more dramatic proportions when we are confronted with the true scale of the lunar landscape. This same place may have seemed ordinary through a telescope, but now you can see it anew. The distance the lunar rover has traveled also will be readily appreciated, as is the vast, lonely, and remote topography. Situated here, could one feel any further from home, planet Earth, or more lost in the vastness of the universe? A terrifying thought, to be sure!

Fig. 6.9. Pluton and *Apollo 15* Lunar Module (photocourtesy of NASA Johnson).

Hopefully, these images will invigorate you to reexplore the Moon, and to visit the other *Apollo* sites in similar fashion. There is nothing quite as enlightening as being able to physically "place" oneself on the lunar surface, as it provides something unique to us as observers. Coupled with the views provided in the field of view of your telescope, the challenge pays excellent dividends with a far greater understanding – as if our feet are actually planted on its surface. Certainly, the visiting astronauts gave us a photographic record that is still unrivaled in scope and variety, and you should take advantage of it.

CHAPTER SEVEN

Instant Imaging
of the Moon

For imaging the Moon, there is at least one absolute fact: *There is simply no method that quite duplicates the sight and visual impact of the live view through the telescope, despite its proximity and any logical expectation for easy imaging of an object so accessible.* Indeed, no image, be it photographic, CCD, CCD video, or web cam, via any telescope on Earth, ever quite seems to equal the stunning real-time presence of the Moon in the field of view. And this applies not only to views seen through grand apertures. The great globe hangs massively in near space, pockmarked at the terminator (the divide between lunar day and night) by a maze of apparently sharply cut formations. This exaggerated relief, of course, is responsible for all of our prior misconceptions of lunar "jaggedness," and it appears in the eyepiece to be almost three dimensional, despite being only two! This is in addition to the diamond sharp and incredibly tiny subtleties of detail that our eyes seem uniquely designed to make out. These special qualities remain illusive in imaging, regardless of the aperture used; even with lesser resolutions of smaller telescopes, all of these visual attributes remain present in the live view! Although the best in recent technology certainly allows us to record the Moon's appearance better than ever before (revealing astonishing detail at times), even such advanced imaging still comes up short! However, we are getting much closer.

If time is not on your side, by examining the various imaging choices available (certainly quite extensive), you will soon see that your options are none too many. So, what is viable for us, with time restraints ever present? The images in this volume were taken using the simplest, most effective imaging method possible (CCD video) that was capable, nevertheless, of producing fine results. Although not quite in the "big league" represented in the best imaging today, the approach enables rapidly produced detailed images and preserves the directness of the viewing experience. As with all types of imaging, only when the conditions are favorable will the results be

A. Cooke, *Make Time for the Stars: Fitting Astronomy into Your Busy Life,*
DOI: 10.1007/978-0-387-89341-9_7, © Springer Science+Business Media, LLC 2009

the most detailed and pleasing, sometimes surprisingly so. Hopefully these images, made under a variety of circumstances, will serve further to inspire you once again to spend a little time with the Moon, even if you are restricted by a busy schedule. You may even wish to try your hand at a little instant lunar imaging.

We should always remember that with photographic or electronic imaging of any type, amateur lunar images typically show the lunar surface as it would appear only in substantially smaller apertures than utilized. Nevertheless, many of the challenges of the past are now greatly diminished, not only because so many sizeable and sophisticated telescopes are now readily available, but also because of the advances of imaging technologies themselves. Let us quickly review what is available.

That old standby, drawing, really has never been in the cards for most people; the Moon is an extremely difficult and tedious subject for such exacting work. Lunar cartography is more like artwork. Despite the prospect of being able to include as much fine detail as may be visible, the fact is that, even when brilliantly executed, it never seems quite to recreate the actual appearance of the surface as well as can even the simplest imaging camera. Worse, it takes an enormous amount of time! Although many observers of the past, and even some today, have produced some remarkable lunar draftsmanship, lunar drawings have become largely obsolete. Unless one is looking for a challenge, or an end result that may be the most truly personal, there are better ways to spend whatever time you have. Not only is the challenge of drawing the Moon from the eyepiece greater than for any other in space, it is also a somewhat different skill; mastering lunar cartography may not necessarily prepare you to draw other celestial objects.

You can attain fairly acceptable images by the oldest way of all, of course: simple photography. However, even when utilizing today's fastest emulsions, unless the exposures are short enough, you are still liable to suffer from the added distortion from atmospheric turbulence, and hence blurring of the detail. The faster the film emulsion, the grainier it may be, too, so speed is not necessarily on your side. Obviously, larger apertures permit the greatest flexibility because their exposures can indeed be kept briefer, but considering the ever-quivering atmosphere you must contend with, the challenge is in knowing exactly when to click the shutter!

Digital cameras and CCD imaging devices offer our best imaging, but they are not the easiest things to master, partly because of multiple complications (especially with CCD), but also for the very same problems as we have always encountered with atmospheric stability in conventional photography.

Frame integrating CCD video cameras and even possibly the "lowly" web cam would seem to be easiest to use – another plus of these systems being the ease of showing the Moon on a monitor to groups of people very effectively, rather than having your guests line up for a brief glimpse through the eyepiece. For the Moon, with optimized streams of exposures, and possibly multiple "stacked" or integrated frames (the compounding of multiple images by several means), the results these cameras are capable of delivering begin to rival their grander true CCD cousins. However, the easiest method of all involves using only single video frames. *Non*frame integrating CCD video cameras, such as the Astrovid 2000, are still useful for lunar imaging to this day, because of their simplicity and remarkable effectiveness. The Astrovid 2000, while not exactly an antique by the commonly accepted term, seems old today since, technologically, things have progressed a fair way in the few years since its making. Nevertheless, frame for frame its quality still remains equal to most new cameras, and its video stream is the same as one made *without* the frame integration common to most CCD video cameras today. One may utilize the most advanced CCD video cameras in exactly the same way. The moving video stream provides a myriad of frames from which to pick the best still image, this being the one that suffers from the least amount of atmospheric distortion. Simply extracting the best possible *single* frame from the multitude is still a great method of lunar imaging; for this, a simple application such as *iMovie* (Apple) works perfectly, with which we can extract appropriate frames.

It is important to fill the chip with maximum information. Although CCD video systems are able to produce some wonderfully illuminated and contrasted views of the Moon at almost any scale, remember that they only reveal their true potential when a telescope is pushed to its limits. We have already explored the reasons for this in Chap. 3. However, we still need extremely steady seeing, especially for the high powers required for revealing optimal detail; needless to say, the kinds of conditions we all dream about do not occur too often. Dark skies are irrelevant, as we could claim quite accurately that the Moon is a worse light polluter than the light from almost any city! We still have to bear in mind that the limits of resolution of the video system itself place an upper limit on the finest detail that can be captured. However, by properly utilizing the available pixels, you might be surprised by just what is attainable, even by the method of using just single frames.

Although the focal ratio of the particular telescope used plays an important role, understand that the term "fast" as used in photography and other imaging is only relevant in that it refers merely to focal ratios; in turn, the smallest (fastest) focal ratios permit the "fastest" exposures simply because the magnification at prime focus is significantly *less* than that of most "slower" optical configurations. This results in a field of view that is better illuminated, and hence an exposure will take less time. So it is all a matter of the specific ratio concerning aperture and focal length, and hence angular size of the field of view. None of this helps us much if we cannot obtain sufficiently high magnifications to reveal detail. Although we may always wish for larger apertures for resolution and power, perhaps an ideal balance for all purposes might be best attained with shorter focal ratio large aperture telescopes. This is because most images will not likely be too great in scale at minimum power; we do, after all, need the flexibility to view at lower powers as well, instead of always going for broke.

Aside from those images provided courtesy of NASA, all of the lunar views in this book were obtained using single, virtually unprocessed frames extracted from a moving CCD video stream. As such, the frames correspond to 1/30-of-a-second snapshots – the exposure time of each frame of the stream. By trying to saturate the CCD chip with as much detail as possible, quite high powers were utilized whenever conditions allowed. This is because the advantages of pixel saturation remain, even when we reproduce the images at a lower scale. The CCD video camera was used in conjunction with either 2× or 5× Barlows; the latter produces extraordinarily high power views, but stable atmospheric conditions will dictate the practicality of its use. These times are none too frequent, despite the theoretical desirability of maximum pixel saturation. Although decent images are possible in less than ideal circum-stances, the loss of the most refined detail is noticeable. In good conditions, it is the

Fig. 7.1. Moretus (single CCD video frame: outstanding conditions). (AC)

resolution of these minute details that separate ever-larger apertures apart from the pack, of course, so we will be trying to capture as many of them as we can. Effective though they are, nevertheless, the images still do not equate with the appearance of the Moon at the eyepiece!

Truly exceptional conditions are memorable occasions indeed, allowing such high-quality images as seen here in the remarkable single frame of Moretus (70.6°S, 5.5°W), Fig. 7.1, not far from the terminator and taken with a TeleVue 5× Powermate Barlow lens. The kind of granular surface texture and refined detail, all too familiar in many live views through larger apertures, jumps out at once from the scene, a clearer hint of those qualities not easily transferred to images.

Resolving Lunar Detail with Digital Video Imaging

The problems of capturing fine lunar detail, by any means, will be immediately apparent, once you try it for yourself. We are still dependent on that old bugaboo, atmospheric steadiness. Do not waste whatever time you have by attempting to make your *best* images when conditions are not as good as they could be; imaging will be no more successful than when you observe in mediocre conditions! Completely optimal conditions are far from the norm, so you probably realize already that only on rare occasions are the highest theoretical resolutions, and best images, possible to obtain. The problems of utilizing a telescope's full potential increase in the larger sizes, because ever-greater wave fronts of air also cross ever-greater apertures. It is important to note that when most of the lunar images were taken for this volume, the air was seldom optimal, though it was by no means poor. However, it does illustrate that most of the time, reaching the highest theoretical resolution potential of larger telescopes will usually dangle unfulfilled before your eyes! Nevertheless, you should still try to obtain all that you can.

Let us start by more thoroughly demonstrating the value of reaching for maximum pixel saturation with the old favorite, the startlingly "clefted" crater, Petavius (25.3°S, 60.4°E). Sir Patrick Moore, referring to it in dramatic fashion in his legendary writings for amateur observers, long ago firmly established this feature as a key object to look for soon after new Moon. This means it is an early evening sight. Although seeing the cleft is relatively easy, optimal observing or imaging of the great crater is frequently affected by air currents and cooling down issues in the early evening. However, because Petavius may also be seen at the other side of the monthly apparition, it is one of the most likely candidates to benefit from observations made *after* the full Moon, when we have the best chance of seeing it well.

Presented here (Fig. 7.2a, b) is Petavius as it appears shortly after the full Moon. The famous cleft clearly stands out. These images were made during the same observing session, in which atmospheric steadiness was fairly good though not exceptional. Both are presented here at the same scale so that easy comparisons may be drawn between the resolution of one image versus the other. Figure 7.2a is certainly not bad when using a 2× Barlow, but in Fig. 7.2b, imaged utilizing a 5× Powermate Barlow, it is noticeably better. We can certainly see more refinement of detail in the latter image, but even better results are yet possible in the best of conditions. Look carefully at both images; although, casually, they may appear much the same, with careful examination the difference is clear.

Going still further, it would be possible to utilize more advanced imaging and processing techniques, such as multiple frame stacking, "unsharp masking," manipulation of contrast, and so forth. The results would be ever-greater clarity

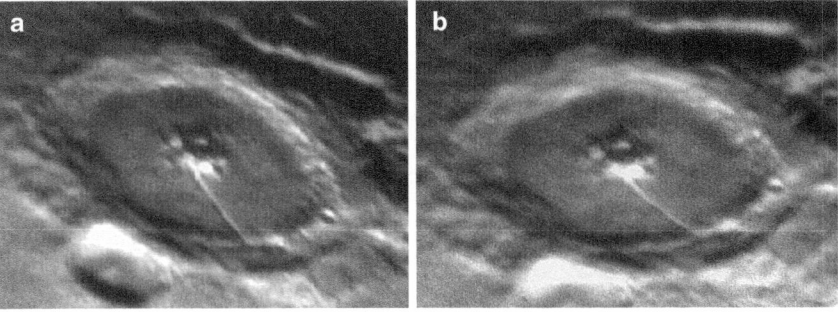

Fig. 7.2. Late evening on Petavius. (AC)

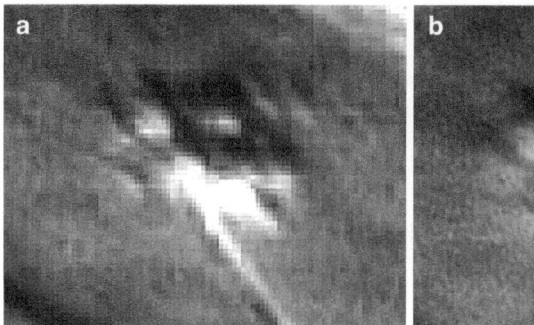
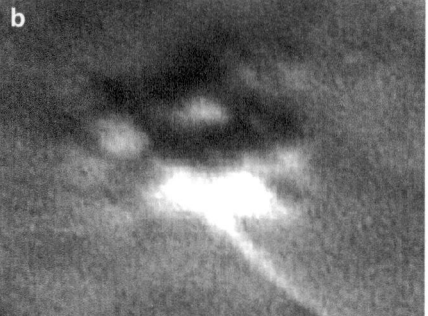

Fig. 7.3. Central peak, Petavius. (AC)

and resolution, more photograph-like quality, and images ever harder to tell apart from those utilizing the most elaborate true CCD equipment. However, as always, going to these kinds of lengths is something many people will not be prepared or able to do from a standpoint of time limitations. Certainly this simple method represents a lot of imaging value for minimal effort.

The effective saturation of CCD chip pixels can easily be demonstrated further, by increasing the scale of the frames to finally reveal the structural grid of the pixels themselves. Here, portions of the two prior frames of Petavius are presented (the central peak of the crater itself), again at identical scales, but now highly magnified. You can readily see the actual pixilation as it begins to separate out. In Fig. 7.3a we have taken things beyond practical value, versus the much reduced degradation shown in Fig. 7.3b, where much more information is present, because less magnification is needed after the exposure.

These images ably demonstrate further how this method can provide a simple, yet extremely effective, means to produce lunar imagery very quickly, with no special techniques other than what is described. Is not this the key to everything we do, when time is at a premium.

Quick Project: Experiment for Effective Pixel Saturation with CCD Video

Time Required: 30 min

First of all, be sure to wait for a reasonably steady night, for best results. Ideally, you should have two or three "stackable" Barlow lenses available, so that you might try different magnifications. Taking the examples from this chapter as a guide, select and record a given lunar feature at a scale according to your judgment. You can almost guarantee that your initial instincts will be too low! Repeat the process with increasing magnification until the image begins to deteriorate, noting the Barlow lens, or combination of lenses, of the last effective video stream. Be sure also to note the viewing conditions; these will be your reference points for the future.

Later, select the best individual frames from the various video streams made at different scales, and adjust them on your computer monitor until they are of comparable size. Note the scale at which you can make out the maximum detail; you will probably be quite surprised how large a scale produces the best results. By all means make further small adjustments (contrast, etc.) to the final image; such minor additional efforts will not take very long.

You will find this to be a very instructive session, which will save you a great deal of time and unsatisfactory efforts later. The bonus will be the stunningly detailed frames made at your most successful image scale – wonderful imagery in a total of mere minutes from start to finish of the entire process, the ultimate in easy lunar imaging.

CHAPTER EIGHT

The Lunar Terminator

Viewing the *full* Moon for the first time through a newly acquired telescope has probably been a huge disappointment to many a budding lunar observer. The novice soon realizes that the most efficient way to gain insights into our neighbor's landscape is to spend time instead at that ever-transitional region of relentlessly moving shadow, where night turns to day, and vice versa, before or after a full Moon. This is the region of the lunar terminator, where lunar observations are most telling and rewarding. It soon becomes evident that most of our lunar observing should take place here, or at least nearby. The dramatically exaggerated appearance of even the most minimally vertical of lunar features is one of the most amazing optical effects you may ever see. The stark elongated shadows, not muted by the presence of an atmosphere, provide dramatic views of the lunar surface.

You already know that the lunar landscape is nothing like the impressions you may have had of jagged and sharp features thrusting high into the surrounding space. Look again at the view of the lunar limb in Fig. 6.1, Chap. 6, and any ongoing impressions about such features will soon vanish! However, by maintaining true perspective it is possible to gain quick and direct insight into our neighboring world, one we are witnessing from such remarkably close proximity. With such detail so startlingly clarified, there is hardly anything we can observe in space that offers so much in so little time. In just half an hour we can see and appreciate more detail on the Moon than we can accomplish in an entire night with any other subject, or group of subjects.

With only moderate apertures, say, 8 in. or more (20 cm) and the relatively high magnifications they allow, you may readily understand why the drama at the terminator makes it hard *still* to fully detach ourselves from the lure the Moon had on many of us not so very long ago. If you find yourself ignoring the truth and simply marveling at that "jagged" surface that you might have once thought actually represented reality it is not such a terrible crime in the great scheme of things. Because of

A. Cooke, *Make Time for the Stars: Fitting Astronomy into Your Busy Life,*
DOI: 10.1007/978-0-387-89341-9_8, © Springer Science+Business Media, LLC 2009

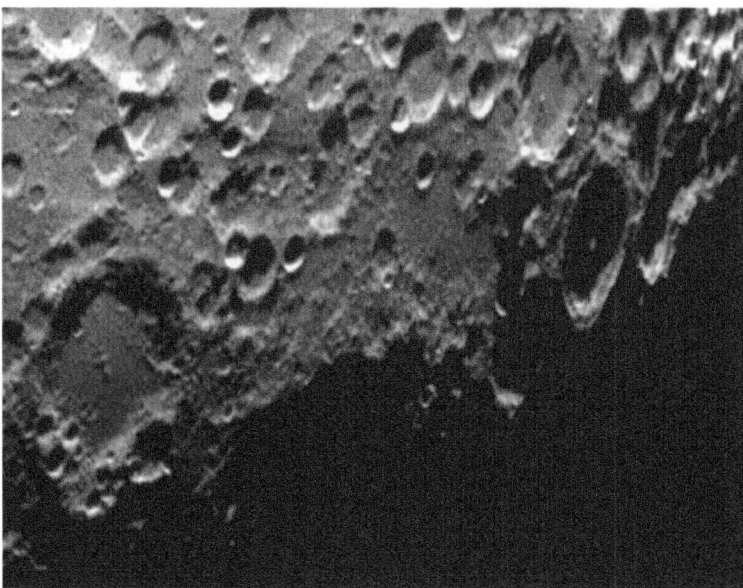

Fig. 8.1. Drama of the terminator. The lunar terminator in the vicinity of Maginus (50°S, 6.2°W: lower right), and Moretus (70.6°S, 5.5°W: upper center) is easily recognizable by its central peak. Mostly in shadow, at lower middle right, is the huge crater Clavius, about to be flooded with sunlight as night becomes day on the Moon (CCD video frame 18-in. reflector).

the illusion of features being larger than life, we are certainly given a feeling that we are closer to the surface. It is also especially remarkable how very differently any part of the lunar terrain appears just a day after being right on the terminator. A few days more, and any suggestion of features being dramatically and toweringly shaped is completely eliminated. Thus, it is all the more surprising how easily and unquestioningly we casually accepted those tremendous exaggerations as being the actual contours of the surface. Most of them, once starkly revealed, eventually disappear completely from view; it turns out that the towering peaks were, in actuality after all, just gentle bumps scattered over a vast expanse (Fig. 8.1).

Quick Project: A Different Way to See the Moon

Time Required: 30 min

Because the new Moon follows the setting Sun so closely, successive phases and the times of Moon rise occur each evening ever later through the course of each monthly apparition. This leads to something else that we may wish to consider, because our telescopes – especially larger ones – may not have had ample time to reach thermal equilibrium before the Moon sets below the horizon. You will waste a lot of valuable time trying to coax a decent performance from your telescope at these times if it has more than a modest aperture.

As alluded to in Chap. 7, in order to properly see certain features in the thin crescent of the new Moon normally associated with the early evening, try viewing them at the other end of the lunar cycle. These features, initially close to the new Moon, are likely the best candidates for viewing after the full Moon. Most observers spend very little time viewing the Moon after it has become full, because of the late observing hours dictated by the waning Moon's appearance in the sky. Even amateur astronomers sometimes like to sleep, after all!

However, after the full Moon the terminator reappears in reverse fashion, with the Sun now illuminating our satellite from the opposite side. The result is that the light and shadow that define the lunar features are now projected in the opposite direction to those that we saw before the full Moon. It is surprising how very differently many of the best known features will now appear, and thus you will likely find an entirely new subject to observe at these later stages of the monthly lunar apparition. In the early hours your telescope will be better able to deliver the kind of performance you long dreamed of!

Quick Project(s): Finding Specific Regions of the Moon and Features at the Terminator

Time Required: However Much You Have!

All of the following types of lunar features, optical effects, and particular examples of destination can be considered as quick observing projects. As such, they do not require specifying beyond this; you can have many hours of cumulative pleasure from preserving just a little awareness of what you are looking at, other than merely perusing the "generic" lunar surface. Conversely, all the types of categories listed through the end of the chapter may be observed in just short sessions, which will always be productive if you are specific in what you are trying to observe.

The Heavily Cratered Midsection

This region shows some of the most rugged and interesting terrain of all. Particularly magnificent is the south polar region during transitions from first to second quarters, and third to fourth quarters.

Shortly after first quarter, the huge and well-defined crater, Clavius (58.4°S, 14.4°W), will immediately draw attention to itself, conspicuously nestled amid a tortured and pockmarked landscape (Fig. 8.2). A more magnificent lunar sight is hard to find. Note its famous inner crescent of ever diminishing inner craters, displayed in a "sickle" crescent.

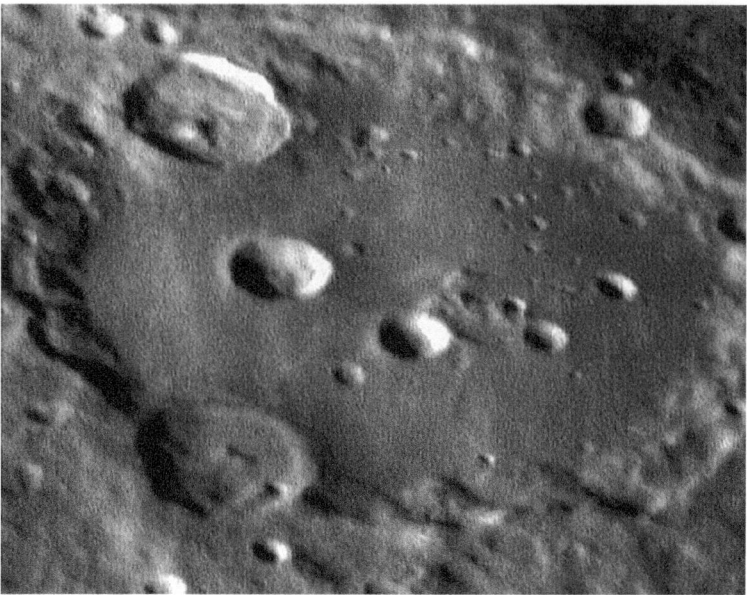

Fig. 8.2. Clavius (CCD video frame 18-in. reflector: outstanding conditions). (AC)

Sunrise and early morning are the most interesting times on the Moon. Aside from the exaggerated effects of height and shape, the absence of atmosphere assures that deep, long, and dark shadows are cast immediately at the terminator, providing our best opportunities to see tiny details in the resulting extreme contrast. As first quarter approaches, the shadows across the center of the Moon's disc reveal some of the most rugged and best of the Moon's features. These are some of your most opportune observing times. Since these are mid-evening and late night sights, they allow our telescopes to reach thermal equilibrium by the time the Moon is placed high enough in the sky to show them to best advantage. The visual drama we find here remains undiminished, even in these more enlightened times of space travel. Lunar sunset is no less dramatic than dawn.

The crater Albategnius (11.2°S, 4.1°E) is such a lunar midsection structure and one of a striking group of three grand craters, the others in the trio being Ptolemaeus and Alphonsus. Featuring complex inner crater formations and rugged walls, Albategneus's inner plain is filled with dark lava, like so many other features on the Moon. Because of the relatively late formation of this plain, it is still remarkably flat and smooth, although barely visible here, being almost lost in the shadows of sunrise. In Fig. 8.3, taken at dawn around the time of first quarter, we can see the starkly contrasted lighting of sunrise. At once, intricate detail is visible in the ramparts surrounding the crater. And what startling relief is visible!

As any lunar observer knows, once the lunar day progresses and the terminator passes by, all of these dramatic terrains will soon be rendered virtually flat and featureless to human eyes.

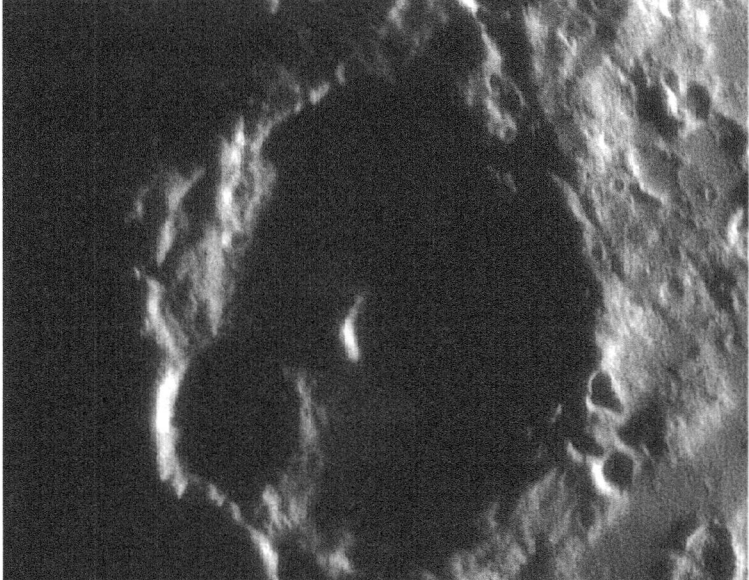

Fig. 8.3. Albategnius (CCD video frame 18-in. reflector). (AC)

Rilles

Lunar rilles are almost as much a staple of lunar observing as are craters and mountains. They are regularly used as tests for telescopic resolution, and indeed, some of the finer ones are very satisfying to observe. These particular features are so numerous on the Moon that you could make a full-time study of them, exclusively.

In a short time spent at the eyepiece, you should be able to derive much satisfaction from observing and counting numerous rilles in any area, especially when you observe one appearing right at, or even apparently beyond, the threshold of visibility for your telescope. Be sure to choose times of maximum shadow for these tiny features, especially the finer ones.

When wide enough and clearly resolved, you can see that the majority of rilles are not V-shaped channels at all, such as would have been the case had they been cut by flowing liquid erosion; rather, their flat bottoms clearly signify another cause. If you look back to the image in Fig. 5.2, Chap. 5 (Prinz region), taken from lunar orbit, the true flat contour common to most lunar rilles may be seen quite clearly. It is believed that the formation of many of them took place at an early date in lunar history, while the crust was still partly molten and lava was still free flowing, the action often actually taking place beneath the hardening surface in the form of molten lava tubes that later collapsed. Many clefts and fine rilles can often be seen within crater walls, and which may have different origins, perhaps the result of crust shrinkage upon cooling.

There are also countless larger rilles crossing vast distances across lunar plains, great in length but often still very narrow in width. In the case of most of them, and unsurprisingly because of their visual delicacy, the only chance to see them is when the terminator passes nearby.

Be sure to spend time with such classic rilles as Hyginus Rille (7.8°N, 6.3°E), the Treisnecker Rilles (5°N, 5°E), or the serpentine Schröter Rille (26°N, 51°W), but also some others, known as tests for optics, such as the fine rille along the floor of the Vales Alpes (49°N, 3°E), or Marius Rille (17°N, 49°W).

Maria

Lunar Maria, those vast lava plains, often appearing near featureless, occupy so much of the surface that we will inevitably find ourselves spending at least some time with them. Any first impressions of featureless terrains, though, will soon dissipate as you become aware that there is, in fact, more of interest in these regions than may be apparent at first glance. Usually darker than the remainder of the surface, lunar maria appear in many ways as would our own seas and oceans as seen from space, which would explain why our ancestors simply assumed them to be similar bodies of water, naming them appropriately.

Again, it is at the time when the terminator crosses these vast regions that we will have the best chance to see much more than flat, featureless expanses of lunar terrain. Formed from lava flows at a later time than most of the more rugged features of the surface, you may observe numerous tiny craterlets within them, created by meteoric impacts that have occurred since their formation, as well as isolated rocky mountainous formations jutting upward, looking for all the world like islands! You will also frequently see highly pronounced rippling effects formed and spread across lunar maria, grand wavelike formations in what was once cooling lava flow.

The noticeable waves and rippling show again that the ancients were even more apt naming these plains "mares!" On the northwest side of the lunar disc lie some of the greatest lava plains of all, and some of our best viewing opportunities will be found here. Many grand craters must have been buried by the heavy volcanic flows of antiquity; many are filled with the same material, Plato being one of the best examples.

Lunar observers have typically taken great pride in counting craterlets within Plato's walls, among others. If this kind of activity appeals to you, then it is certainly an activity that could be repeated for ever-greater crater totals during short observing sessions, and within any structure or complete mare.

Perhaps the most famous, the Mare Crisium (The Sea of Crises) appears close to the limb at the new Moon and thus is in view until well into the third quarter. Dark and easily visible to the naked eye despite its relatively compact dimensions, it is so striking and visible for such a large proportion of each lunar apparition that almost everyone is familiar with it as part of "the man in the Moon," even if they have never looked through a telescope or even known its name.

Ray Craters

No discussion on the Moon would be complete without some reference to the famous "ray" craters. Becoming most prominent at the full Moon, they benefit from flat illumination after the terminator has come and gone, the longer amount of time afterward the better. In the context of our discussions here, it would not be unreasonable to term these spectacular phenomena perhaps as those of the "antiterminator."

Shortly after the lunar dawn as the shadow of the terminator makes its way across the lunar landscape, there is little sense that certain craters among the thousands spattered all across the surface will ultimately take on a different appearance to that of any other. However, as the day grows longer and the three-dimensional relief of the crater walls vanishes into the proverbial ether, we become aware that something else is indeed evolving: the regions surrounding certain prominent craters begin to resemble ray formations, not unlike starbursts in appearance, that often extend for hundreds of miles in all directions. The composition of these "rays" is, in fact, ejected matter from right out of the craters themselves, splashed and spread far and wide by the force of the impact of the bombardment that created them in the early years of the solar system. Because this ejected material consists of a different type of lunar matter, coming instead from beneath the surface, it reflects sunlight differently, becoming most prominent under full overhead illumination, while being almost invisible early in the lunar day.

The most famous of the ray craters is Tycho (43.3°S, 12.2°W), featuring the most striking and widely spread ray system of all. Initially it appears as a prominent and well-formed crater with a central peak. But as the full Moon evolves, the surrounding terrain takes on a remarkably streaked appearance, as brilliant ray pigmentations begin to spread, almost as if glowing in all directions, while Tycho increasingly loses much of its visible structural detail. The dramatic transformation occurring right in front of our eyes is remarkable (Fig. 8.4).

Tycho may be the most celebrated example, but there are other fine lunar ray craters that are well worth your time to explore. These include Copernicus (9.7°N, 20°W), Aristarchus (23.7°N, 47.4°W), Byrgius (24.7°S, 65.3°W), Proclus (16.1°N, 46.8°E), and the diminutive but remarkable Messier (1.9°S, 47.6°E). In truth, you can find traces of *rays* around many craters, but clearly only some of them are sufficiently prominent to be termed thus.

Although the Moon is the only object in space that allows us such ready ease of access – so familiar to us that it might be almost part of planet Earth – it is surprising how much we still do not know about our satellite. Sadly, not too much serious lunar research is being carried out by the amateur these days. It would be comparable to thinking that we know a particular region of our own world only from photographs and articles we have seen in magazines.

It is still sad that NASA was compelled to let the Moon go when we had just begun our manned explorations on the surface; visionaries have always been in the minority. In any event, the truth is that international politics and superpower rivalry had more to do with the actual quest for the Moon than any purely esoteric impulse. Apparently, however, the not too far distant future promises to serve us better with newly announced plans for establishing a lunar base by 2020. That still seems like a long time compared with all that was promised not so very long ago; it was not considered completely unrealistic when, in the 1970s, NASA proposed sending people to Mars by 1984!

For the busy amateur, the Moon remains unique in the simplicity of approach it provides. It serves doubly in that it is an excellent starting place for any newcomer to astronomy, because it is so gratifying to have such immediate and exciting results when viewing another world across space. The Moon can still be used to begin training

Fig. 8.4. Tycho (CCD video frame 18-in. reflector). (AC)

and to refine the eye's astronomical observational skills, without ever being a source of frustration. We should never allow ourselves to become blasé by its willing compliance. If you are fortunate enough to own or have access to a moderate to large amateur telescope, you already know how completely overwhelming it is to view the lunar surface through such means. For lunar phases, an invaluable service on the web for rising and setting times, as well as phase sizes, and much more, see Chap. 16 in this book.

Hopefully, reading the last three chapters will have persuaded you to look again at the Moon, should you have become jaded by its familiarity. There are revelations to be observed and studied in almost every segment of the Moon and throughout its varying topography. Different degrees of shadow reveal all manner of things, which was never the intention to cover here, this being the stuff for your own explorations. By making the mistake of simply dismissing this most familiar of all space objects as an object whose time has now passed, you will miss some of the best and most easily enjoyed times with a telescope. It is time to rediscover all that our partner in space offers. With time being precious, we cannot afford to ignore such compelling material. Make time for the Moon, too!

Section III
The Greater Solar System

CHAPTER NINE

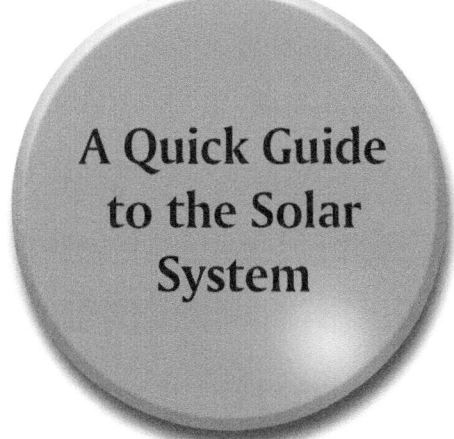

A Quick Guide to the Solar System

The solar system still remains a defining part of many an amateur astronomer's interest in the universe. These days, however, actual viewing of these subjects is being subjugated increasingly to the more indirect experience of CCD imaging. Effectively, this has taken much of the human element, and more particularly, the ready accessibility of these subjects, out of the equation. Some spectacular close-up views from spacecraft also may have robbed amateur observers of some of the wonder they used to have.

However, it is not all over for the visual astronomer. We should never fail to make distinctions between live telescopic viewing, telescopic imaging, and scenes imaged by spacecraft on the surface of the planets themselves. These all remain distinctly different types of revelation, and they are all valid and valuable. However, the direct, live experience remains not only the most personally impacting, but in a busy world it is still the simplest and quickest way to access our celestial neighborhood. There is still nothing quite like the crisp refinement and breathtaking impact of the live view, something impossible to duplicate by any form of imaging, and certainly far less time-consuming.

The good news, therefore, is that our old solar system favorites can remain the core of our astronomy, if we so desire. Take advantage of the opportunities you have here, because once we leave our neighborhood and enter deep space it is no longer possible to have such intimate contact with anything. We still have no inkling at all how any destination in the greater universe would actually appear close-up in its *own* neighborhood! Thus, the entire realm of what has always been the amateurs' solar system is still meaningful in the most dramatic way. Regarding any thoughts we might have about participating in a useful scientific role, we must accept that

A. Cooke, *Make Time for the Stars: Fitting Astronomy into Your Busy Life,*
DOI: 10.1007/978-0-387-89341-9_9, © Springer Science+Business Media, LLC 2009

while some forms of specific, if limited, continuing surveillance are indeed still feasible (such as that performed by networks of amateur organizations), in many of the traditional categories of observing, contributing in such a capacity is largely over for us. Regardless, the wonder of spending time in our astronomical backyard remains undiminished, but perhaps being mere sightseers serves to free us from the burdens and demands of being more serious students.

Although the new perspectives provided by visiting spacecraft have made the solar system a much more familiar place, we cannot pretend that this has had no effect on our own objectivity; it does, in fact, cause us to see things in an entirely different light. We should not ignore these insights. Although this does indeed make for a certain contradiction, it nevertheless provides the kind of reality we used to only dream of having, and one we cannot pretend does not exist. Spend a little time looking at the multitudes of images provided at the NASA web (see Chap. 16 "Astronomy via the Internet"), as they will provide unique perspectives. These sites are updated continuously, and between them feature the full range of space missions since the beginning of the Space Age, as well as extensive Caltech-related observatory imagery and descriptions. You will see that this goes far beyond merely the solar system!

Solar system viewing is a means for developing refined viewing skills, which will help you greatly in other areas of astronomy, too. It has to be said that developed visual abilities fuel a special appreciation all of their own, for it is those glimpses beyond the normal visual threshold where the greatest rewards lie. For our observations, we have already covered some of the most useful equipment, along with new and better products. However, despite all the apparent advantages offered by today's high-quality equipment, many observers are spending less time at the eyepiece! This is not because they have found how to accomplish more in less time, but rather they are using the telescope merely as a necessary interface to the computer screen. Not only does this destroy the entire live viewing experience but also virtually guarantees that fitting astronomy around a busy personal schedule will become impossible. (If you have already tried to do anything quickly on a computer you are probably already well familiar with this scenario!)

The Sun

The dominant force in our own existence, the giver and the taker of all things physical in our own realm, the only star near enough for detailed study, and the center of everything in this realm, is the mighty Sun (Fig. 9.1). Nevertheless, it is the one object that many observers remain uncomfortable exploring to any degree firsthand! Part of this "solar phobia" stems from earlier years when we did not have access to modern protective filters and other such modern equipment for our telescopes. A healthy respect for the Sun was drummed into us by all that we read and heard. Despite plucking up enough courage for the occasional peek or short session, many observers still find that they cannot overcome the fear that a compromised filter, Mylar film, a scratched glass objective shield, or the shattering of glass of any kind would bring their worst nightmares to life. If you do not share in these misgivings, you may

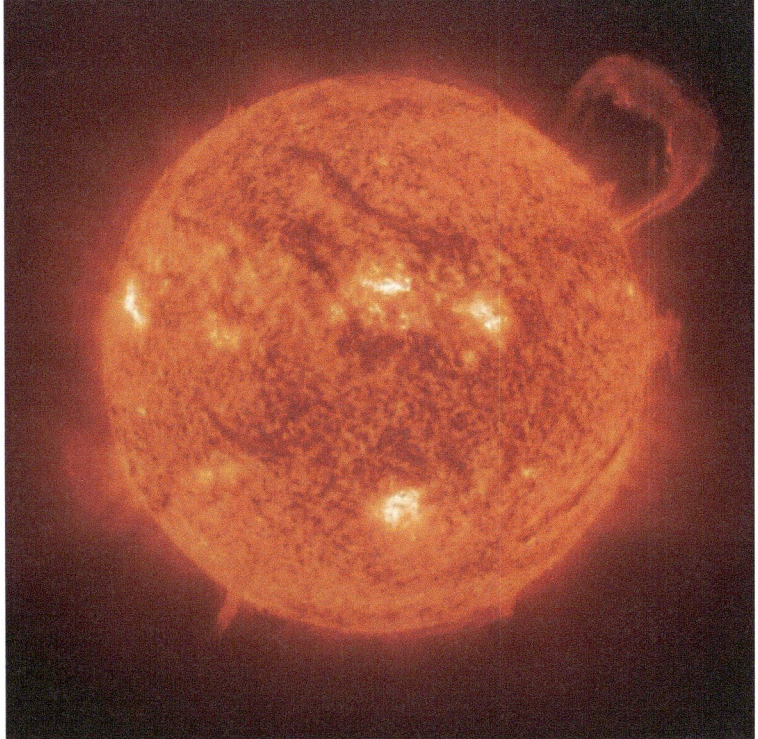

Fig. 9.1. The Sun, SOHO image 1999, Extreme Ultraviolet Imaging Telescope Consortium (photo courtesy of NASA/JPL-Caltech).

find the Sun to be among the most compelling subjects in which to immerse yourself, and it will certainly deliver a great deal for limited time, opportunities, or constrained budgets.

Being a unique subject (even the *next* nearest star lies at an almost unfathomable distance), obviously the Sun is prime fodder for study, and it is study that offers easy and generous access for busy lives. In having this unique opportunity to study an actual star close-up, it is a natural destination for the observer. A complete subject all unto itself, the Sun is worthy of an entire separate volume; indeed, some observers never concern themselves with anything else. Many of these enthusiasts spend their annual vacations traveling to far away places to observe total eclipses and will accept remarkable degrees of discomfort, expense, or inconvenience to do so!

Even today, many misconceptions still exist among average laymen regarding the most everyday fundamentals about the Sun, such as "it is a giant gas ball burning in space," when, in fact, its heat and light is the result of a gigantic nuclear reaction. Many people do not even realize that the Sun is a star, and not a particularly special one at that. Even more naively, expectations that its output will remain forever at a constant amount, or that if it changes at all during anyone's lifetime this is abnormal, have lead the average layman (even many in the scientific community) to believe that humankind alone is responsible for apparent global climate change, the "hot" topic of our time. To discount the probable influence of the Sun, apparently a variable star, is shortsighted, to say the very least.

The misconceptions do not stop there, however. Many amateur astronomers, although easily able to explain such simple things as the Sun's movement across the sky, might be hard pressed to explain to the uninitiated why it appears to change direction in the opposing hemisphere, and more to the point, when exactly does this apparent decisive turnabout occur? Once we cross the equator, is there a sudden reverse in the Sun's direction across the sky? In case you, too, find yourself scratching your head, just visualize the position of the Sun as it gains altitude in the sky as we approach the equator. We are still turning in the same direction, regardless of which hemisphere we are in. Once we move ever more into the opposite hemisphere, it will appear to have crossed the highest point in the sky and drop lower into the sky in an arc opposite to that from which it was traveling before. We will now find ourselves *turning around* to view it because of its placement, and it will now appear to be moving in the opposite direction, even though only its relative placement in the sky has changed! Logically, we perceive the Sun's 'movement' by *facing* the arc of the sky in which it appears to travel, from our home base or any other. All in all, this is a pleasingly simple explanation for something that often leaves otherwise well-informed folks stumbling for words or clear mental visualization. There are, of course, many other examples of a similar nature that we will not cite here.

We cannot complain that insufficient light or too small an image scale is a problem with the Sun! An effective study of it may be undertaken, therefore, with relatively small telescopes. Even with fairly limited apertures and appropriate filters, some solar devotees have spotted remarkable amounts of Sunspot activity and detail. Even an untrained eye can see far more detail on the Sun more easily than almost anything else in space. This is not to say that grander results may not be obtained with ever-grander equipment. Indeed, a large, unsilvered primary mirror will provide miraculous projected views, but few observers will be prepared to equip their large telescopes with unsilvered optics, or to have another set of aluminized optics on hand for the rest of their viewing. The need for such measures is because the heat generated by the Sun via large highly reflective standard objectives would invite equipment catastrophe. And even such special mirrors may be subject to disaster. Large filters for the primary optics are neither sufficiently effective, reliable, or even available. Some dedicated solar telescopes made by amateurs project an enlarged image of the Sun onto a flat screen; many serious solar observers prefer these for their observing. However, perhaps an ideal situation is to use specially designed solar refractors (such as those originally made by Colorado Instruments). They would seem to be the easiest and best "Sun scopes" of all, and certainly best for enthusiasts without much time on their hands.

John Watson, formerly editor of astronomy at Springer, has made a substantial contribution to this book on solar observing and imaging in Chap. 13, "Daytime Astronomy."

Rediscovering the Planets

Even more than with the Moon, it had been all too easy in recent years to dismiss the planets as astronomical "has-beens," whose interest and meaning for us had been left behind in the wake of modern cosmology and the far-flung universe. Some of us never bought into such a status for the solar system, however, and continued to observe these traditional astronomical sights with undimmed enthusiasm. However, a sudden new energy in official scientific circles has come about now for the planets, due to the astonishing discoveries made by recent unmanned spacecraft. The direct ties that many scientists now see in the relationship between the solar system, the universe as a whole, and the formation of life here on planet Earth itself has given professional astronomers reason once again to return to these relatively nearby *has-beens*. More amateurs are returning, too, while some of us never left.

For amateur observers, three grand destinations form the mainstay of our observable planetary system, that is, Mars, Jupiter, and Saturn. These are ideal telescopic objects, and for many of us these three planets, and/or the Moon, were the reason we entered the hobby in the first place. Although there is something quite mystical about all of the planets' place in space, it was these three worlds in particular that kept many observers straining to extract complex detail, much of which was as yet unconfirmed, not understood, or even unknown. We could only speculate about the actual nature of much of what we were seeing, and we could only imagine far away fanciful landscapes.

However, there is another, less immediately obvious, consideration for spending time in this corner of the universe. The fact that we can see these particular objects with such refinement, in real time, and *in full color* is more significant than it may seem. Color is a rarity once we leave the solar system. It is not that it is not there; however, experiencing it to any noticeable degree at the eyepiece is usually not in the cards. Now, aside from the brilliant colors of many individual stars, which are often striking to be sure, we are referring to almost all of the much fainter large structures far outside the solar system - especially outside our galaxy - where all significant traces of color are largely lost on our eyes. At these very low light levels our eyes strain to register anything at all, let alone color; those few objects that do reveal it present only the vaguest suggestions in faint pastels, at best. So take advantage of all that is nearby!

The Use of Color Filters

Many things sound possible, even promised in all the descriptions we see about filters. In theory, at least, the concept of blocking parts of the viewing spectrum with the appropriate filter to emphasize details of coincidental spectral properties certainly sounds good. It seems logical that it should work, and sometimes to a small degree it actually does, although it is unlikely to come close to the level of anticipation you may have. You will find that, with just a little care, you can see the same details *without their use*, albeit sometimes a little less starkly, providing a far more natural and satisfying state of appearance. Sometimes filters succeed only in coloring whatever we are looking at! They will not magically transform any view and their use often results in something far less desirable than using no filter at all. And, oh yes, they usually produce enough "ghosting" as to ruin the performance of any good eyepiece! So, just as the practice of stopping down the aperture of a telescope is a questionable practice for improvements in planetary resolution, you may generally find that the live filtered view generally loses more than it gains.

Should you wish to explore this topic further, including finding possible resources on the Internet, see Chap. 16. The Internet resources should provide some helpful and detailed views on many specific filters from the standpoint of a true filter "disciple." There is always the possibility that perhaps, correctly applied, they might work for you! Meanwhile, you will need to take into consideration the aperture of your telescope, because too dark a filter for some apertures will be just right for others, and vice versa. Various shades of green, even yellow, are widely held to enhance detail on Jupiter's equatorial zones, and paler varieties are claimed to provide specific enhancements of details on Uranus and Neptune. However, the slight and vague nature of the markings on these outer solar system subjects must be understood, together with the unlikelihood that you will never see detail revealed on their surfaces! For the inner planets, a violet filter certainly is useful for discerning details of the cloud system of Venus, although there still will not be much of that to be seen either, regardless! So always be realistic in what you expect any color filter to do.

Everything Else in the Solar Realm

Perhaps surprisingly, the planets inside Earth's orbit are much less accessible in terms of viewing than those outside. Searing hot inner planets Mercury and Venus circle the Sun so closely that their temperatures long ago eliminated the prospect of finding anything on them other than parched landscapes. Venus, because of the thick cloud layer enveloping it, and Mercury, with its greater distance, poor viewing placement, and diminutive size, seem to have always conspired to yield little detail through our telescopes. Fascinating morsels of imagery on their surfaces itself via spacecraft have unlocked some of the secrets of Venus's cloud shrouded surface, the hostility of this world confirmed in dramatic fashion. Infrared imagery has also shown us the entire planet's terrain, albeit less satisfyingly than if we could just get rid of the clouds and see it for ourselves! Because of its total lack of atmosphere, Mercury has been fairly well imaged from near its surface by unmanned probes. Suffice it to say, the surfaces of these two planets remain places upon which no human may ever set foot.

In the opposite extreme, far beyond the orbit of Saturn, far-flung Uranus and Neptune are only just beginning to give up their secrets, thanks again to space probes. However, again, just as with Venus and Mercury, it is likely that amateur observers will find these planets somewhat frustrating to study. Their great distances prevent amateur observers from seeing anything other than faint single-colored discs.

Further yet, beyond Uranus and Neptune but still within our solar system reside numerous additional worlds, with new classifications that have become a red-hot issue within the scientific community. Minor planets and asteroids will not reveal any outline or detail, of course, but many amateurs find great satisfaction in searching arduously for the brighter members of the Kuiper Belt fraternity; image intensifiers may be of value here since we are seeking points of light, which are generally well shown in such viewing.

We must not forget comets, asteroids, and meteors; these are all part of the Sun's realm. For some observers, these subjects alone are sufficient to occupy all of their attention, although if your time is limited the requirements are such that they will probably eliminate any serious depth of study. Certainly, the mystical presence of a bright comet in the sky is something in which all people, astronomically inclined or not, will take interest. However, the study of these celestial visitors is often most effective with large binoculars. Leftover debris of various solar system events of antiquity, meteors also respond very favorably to image intensifiers, since they generate wide spectrums of light as they burn up in Earth's atmosphere. The various annual bright meteor showers always attract a wide assortment of devotees, who find great satisfaction in logging the number that fall in each hour. Most people, though, will find their biggest pleasure in merely watching the magnificent cosmic spectacles, nature's own fireworks displays, from a reclining chair. As always, dark, transparent skies are best, and you are almost always guaranteed to see at least one dazzling meteor, which may appear more as a fireball, during any night out in these conditions.

Occultations of stars by planets are also sources of great anticipation, although any chance of seeing a bright star eclipsed by a planet is rather rare. Many amateurs derive much interest recording the effect that a planet's atmosphere has on the light source by the speed in which it is snuffed out. It is possible to draw some important conclusions from the results, which may even have further scientific value, although this is another rather work-intensive occupation that may be of little practical value for you. It is a more common occurrence for the Moon to occult something bright, because of its greater size, and of course, the Moon has no atmosphere. For most of us, however, the result is more spectacle than anything else, because stars will disappear instantly; planets will do so a little slower, of course, because they have appreciable dimensions from our point of view. However, events such as these are still quite rare, and this highly specific viewing activity only occasionally develops into a fulfilling area of specialized study for someone. Needless to say again, for such specialized forms of observing, you may simply not have the time.

On Being Useful

These days, so much detailed and complex information concerning the solar system has been uncovered by advanced research that there is not much left to be contributed by the amateur, except in one area. Around the clock surveillance of the three most easily observed planets still does have a certain value, as there is as-of-yet no continually orbiting spacecraft or large terrestrial telescope capable of the full-time monitoring of every large climatic or other ongoing global aspect of these worlds. This leaves certain monitoring opportunities still within the field of study for the amateur observer. Again, though, this will necessitate more dedication of time than you may have. However, it is not always necessary to feel useful, as the pure enjoyment of gazing upon other worlds is always wide open to us. It is this wonderful ingredient that often gets forgotten, as some amateurs seem to always be looking for justification for the time they spend with their hobby. Just reveling in all that there is to see, and making the effort to be knowledgeable about what meets our eyes, ought to be justification in itself.

Now let us look more closely at our local solar system destinations. There is so much pleasure in watching and studying these worlds that we cannot help but ponder the reality that people will actually visit, or even colonize, some of them during the present century – especially Mars! If you are of the younger generation, then maybe one of these people will be you! Meanwhile, all of us can visit them in our own way and come to feel that we know them, even if we cannot actually go there. You already know that there is a wealth of frequently untapped potential for your personal explorations at the eyepiece beyond merely adding to the near endless parade of mediocre CCD planetary imagery. The best part is that much enjoyment may be obtained without a huge time commitment or imposition on your life.

CHAPTER TEN

Planetary Imaging on a Time Budget

A good reason to never overlook our own little corner of the universe is that we can see many of its places of interest so readily and in such detail. We should never forget that within the solar system are destinations we can "visit" and comprehend in "Earth terms." They are close enough to allow us at least a small idea of their true nature, in full color, and in relative vividness at that! Furthermore, we can do it in real time; visual astronomy offers immediacy, which ties significantly into the thrust of this book.

Nevertheless, you may wish to produce some imagery of what you have seen and so might be interested in some of the methods presented here. There is also a need to illustrate what can be seen at the eyepiece, which was the underlying and primary reason for their inclusion. The planetary images in this book were made more as a result of time at the eyepiece than any particular passion to spend it in front of a computer screen or in a photographic darkroom. The fact that effective imagery may be obtained very quickly and easily, and certainly without huge technical expertise, makes the various simple approaches outlined here even more applicable to our purposes. It would make little sense in a volume such as this to recommend that you spend the countless hours required to take part effectively in the standard types of imaging used by many enthusiasts today.

In the past, apart from recollections taken from live viewing, the limitations of the technology ensured that no telescope, professional or amateur, could really give us much for the record in imagery from the planetary realm. The best photographic views formerly provided by such monumental facilities as the Mount Palomar 200-inch look quite blurred and meager compared with the best CDD images produced by many amateurs today, sometimes with apertures of no more than 8 or 10 in.! Those limitations ensured that all things planetary remained primarily visual, and our telescopes were used as direct extensions of our eyes. Today, of course, we now have many options. Because technology has provided the amateur with equipment

A. Cooke, *Make Time for the Stars: Fitting Astronomy into Your Busy Life,*
DOI: 10.1007/978-0-387-89341-9_10, © Springer Science+Business Media, LLC 2009

capable of recording far superior images than those made at even professional observatories of the past, this new capability must be a huge lure to many enthusiasts. However, with that lure comes the possible (and the all too common!) downside in the creation of "manufactured" images, processed almost to the point of free license instead of realism. When the dust settles after all the excitement, however, the nagging fact still remains that very few images produced by any method are able to approach the unique appearance of any planet as seen live through an eyepiece; nothing quite captures the luminescence, the beautiful clarity, and refined subtleties that the eye alone can register. And the lack of awareness of that very difference may be slowly sending the visual approach into obscurity.

With any form of imaging, in the interests of realism one should be keenly aware of the need to keep contrast and color to a minimum. It is frequently common among amateur draftsmen and even CCD imagers to represent these things far too blatantly to appear lifelike. In typical amateur images and drawings, colors and contrast, among other things, usually tend to be far too pronounced; features are represented as if they stand out starkly and obviously in the view. This is possibly a by-product of the users' insufficient time at the eyepiece, but possibly it also may be a result of the observers' *developed* viewing skills! Is this a contradiction, perhaps? It depends. Such skills can make these things clearer to the observer beyond the actual natural appearance. It is easy to forget that what we sketch, or process by computer, may be the result of extended time at the eyepiece, during which the eye and mind have come to discern what is there. It is not uncommon for the brain to convert fleeting impressions into blatant mental imagery that is well beyond reality. At the other end of the scale, the people at NASA are in the habit of *deliberately* overemphasizing the colors of their close-up planetary images to capture the nuances and the continuing processes taking place on the planets. However, color is presented to the public in such overwhelming amounts that it defeats the purpose and changes their expectations. When their turn comes at the eyepiece, a live, telescopic view may well be a big let down!

Drawing

In the quest for realism in this day and age, just raising as a potential method the ancient art of drawing must seem archaic to some, as it certainly seems far removed from the hi-tech approach we now see being relentlessly applied to everything. So just why would anyone make any effort to produce something ultimately more approximate than the exact mechanical representation of a fine CCD or CCD video image? Why indeed, unless we can gain some advantages. However, on the most basic and practical level, it is a very simple and fast way to proceed, once you have mastered the basic skills. Another advantage happens to be for the training of astronomical vision. Because drawing forces us to make definitive judgments in order to put them on the page, you must organize what you may otherwise tend to see only casually; unspecific viewing may reveal almost nothing valuable at all.

Thus, drawing the planets remains valid, despite its limitations and subjectivity, especially when trying to represent as accurately on the page as possible the true appearance of the subject in the eyepiece. For this reason, you might reject, for example, such standard techniques of "dotting in" the borders of clouds on Mars and other such methods used by many amateurs to represent planetary features. If realism, or more precisely, capturing the essence of the live view, is your first consideration, you will probably also elect to place the planetary image against a black background on the page; you would be surprised how significant an effect this has on one's perceptions.

The ultimate result of developed viewing skills is that you end up schooling yourself in a form, one could say, of forced scrutiny, a little bit like learning to see again. Is not this, perhaps, the name of the game you are trying to play – that of extracting the maximum results for the time you have available in the most efficient manner? By drawing you learn to see what is present in the live view. It is also how you become able to discern great amounts of visual information in a very short time, something invaluable to you when time is short. And another benefit is that drawing can produce closer representations of the observational experience itself, more like the impression that comes across to your eyes, even if less technically accurate.

The development of good drawing skills also involves the art of what you might term "controlled staring." This amounts to being able to hold your focus on something small and tenuous, while keeping your perceptions always open and ready to receive. You will soon find that the eye continuously tries to scan the field of view, often reluctantly holding onto the subject. In experiencing this annoying challenge, even as a distinct "flickering" of the eye that seems reluctant to settle, it is because in normal vision such involuntary and unconscious scanning is how we put together composite views. It is important to realize that there are probably no accomplished observers of note who have *not* spent many hours at the eyepiece with pencil and paper, learning the skills required to truly see what is there. Although you may not have many hours, this does serve to illustrate the point. And over a period of many years, you *will* probably accumulate many hours!

The simplest drawings can be made in black-and-white (using a lead pencil), which offers a simpler and quicker method than drawing in color, though naturally less effective in conveying the essence of what we have seen. Still valuable in providing rapid results and eye training, decent black-and-white imagery may be accomplished in very little time. It is relatively easy to achieve a wide range of shading by varying the pencil pressure, but more important, subtleties may easily be obtained by using the blending action of the fingertips.

Quick Project: Drawing Jupiter in Pencil from Observation

Time Required: 10 min or Less per Drawing

Make up a cardboard template (for the slightly oblate shape of Jupiter, trace an actual planetary image), cut it out, and create blank disc outlines on a white page in a sketchbook. On one of the blanks you have made rapidly sketch what you see at the eyepiece; you will notice that disc rotation is noticeable in a remarkably short amount of time, so put as much information – even notes – onto the page as quickly as possible. Allow for new detail rotating into view by making sure you have as much information as possible first toward both sides of the disc. Then move onto the preceding edge, then the middle region, and finally the detail originally at the following edge of the disc – now better placed to complete. New detail appearing should be discounted, because it was not there at the beginning of the drawing.

It is easy to finish the image later. Almost as simple is to blacken the surrounding "space" around a traced disc blank for greater realism, and make multiple photocopies for all your drawing blanks. Proceed by filling in and blending the details.

The examples below of such imagery from the author's sketch book demonstrate several things:

1. That such simple black-and-white drawings still have their place, especially since making them, start to finish, may be undertaken in a matter of minutes.
2. That they are less effective against white backgrounds.
3. And that development in viewing and drawing skills may be seen here in these sketches quite early on, showing the potential possible (Fig. 10.1).

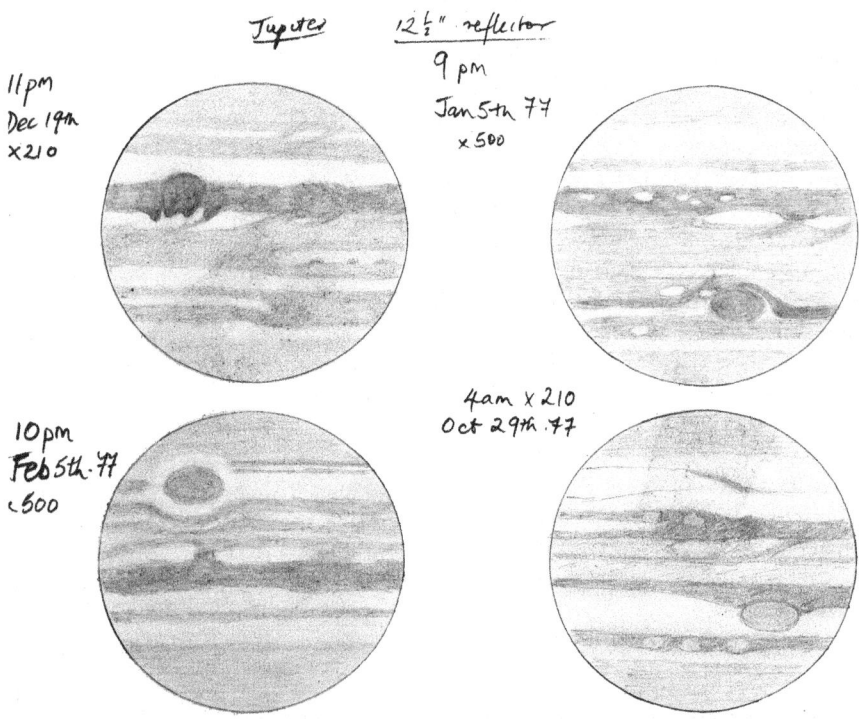

Fig. 10.1. Jupiter December 14, 1976 to October 29, 1977. 12½-in. Newtonian reflector 210x - 500x (You will notice that I was less than particular in maintaining any consistency of polar orientation in those days; the Great Red Spot appears randomly to jump from the northern to the southern hemisphere!).

Quick Project: Drawing Mars in Pencil from Observation

Time Required: 15 min or Less per Drawing

Assuming that Mars is in the sky at the time you undertake this, it is actually quite different from drawing Jupiter. The image is also usually much smaller (requiring greater magnifications), with fleeting details, less obvious contrast, phases, blinding brightness, and many other aspects and potential transitional events (such as dust storms) that conspire to make the challenge somewhat greater. The rotation of the disc, although not as fast as Jupiter's, is nevertheless enough that you will need to apply similar principles.

A lack of understanding all that could be seen on this notoriously fickle object is quite obvious in these drawings, even with the advantage of a good 12½-in. reflector! However, over even just a few short sessions, Mars will begin to yield its secrets, especially if you keep for reference a chart of its surface nearby. Since Mars appears in phase so much of the time, you can either make several cut out blanks of those shapes, or less effectively, simply shade in the disc, as in the first example below (Fig. 10.2).

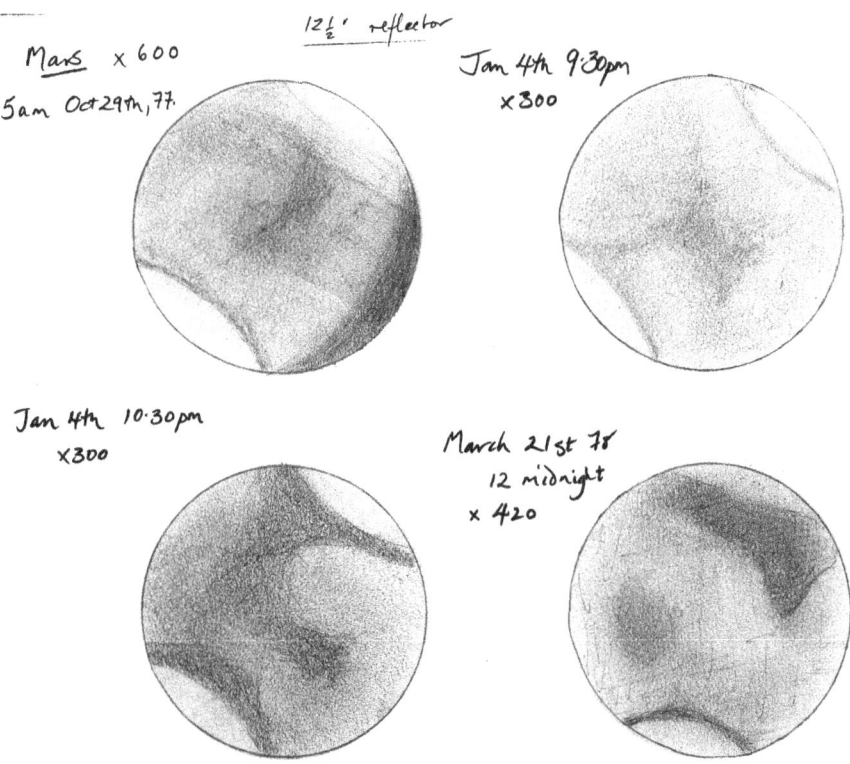

Fig. 10.2. Mars October 29, 1977 to March 21, 1978. 12½-in. Newtonian reflector 300x - 600x. (AC)

Despite the obvious difficulty in making sense of the Martian surface, the Syrtis Major is clearly apparent in both of the lower images, and both polar caps with their surrounding darkened borders show well in all of them; their size reveals that these drawings were made when the planet was still experiencing winter. The vagueness of ready detail indicates that it was one of the least favorable oppositions, with a maximum disc size of only 14.3 min of arc at opposition during January 1978. Perhaps it is the Mare Acidalium that appears in the upper right image, and a trace of it to the right in the image at lower left.

Quick Project: Drawing Saturn in Pencil from Observation

Time Required: 20–30 min

Now this subject provides a much greater challenge!

The matter of proper proportions of the disc itself is as much of an issue as is the correct appearance of the rings, which may be tackled freehand. Because they are in a constant state of change, preparing different templates would likely be advisable if you want to make many drawings quickly. However, if you take advantage of images of Saturn's current ring status, which are available at all times on many Internet sites, you can readily print one at whatever scale you select and prepare your template from that. It is relatively easy to make multiple blanks with black backgrounds, just as you may have done with Jupiter and Mars. If you can detect the inner Crepe Ring, the easiest option may be to omit it from the template, but allow for its width, applying pencil in light shading. Lead pencil drawing still retains the advantage of relatively quick and simple imaging.

These two examples show strikingly what was possible to see and draw *with only 3 in. of aperture* (Fig. 10.3).

Drawing in Color

One of the reasons that color drawing is not the easiest method to master is because of the nature of colored pencils themselves. Color compounds are not nearly so flexible for recording subtleties and nuances as is pencil "lead," so we have to learn how to manipulate a less flexible medium. Color also requires that you be even more conscious of what comprises the details you see. Nevertheless, compared with most of the present-day imaging methods, it remains a simple and fast procedure by comparison. The simple full disc color drawing of Jupiter later, Fig. 10.4, conveys quite effectively the appearance and resolution of its appearance through the eyepiece of moderate to large amateur telescopes. Note the pastel shades! You will find Jupiter in particular a most willing object, always full of variety, colors, and plentiful detail;

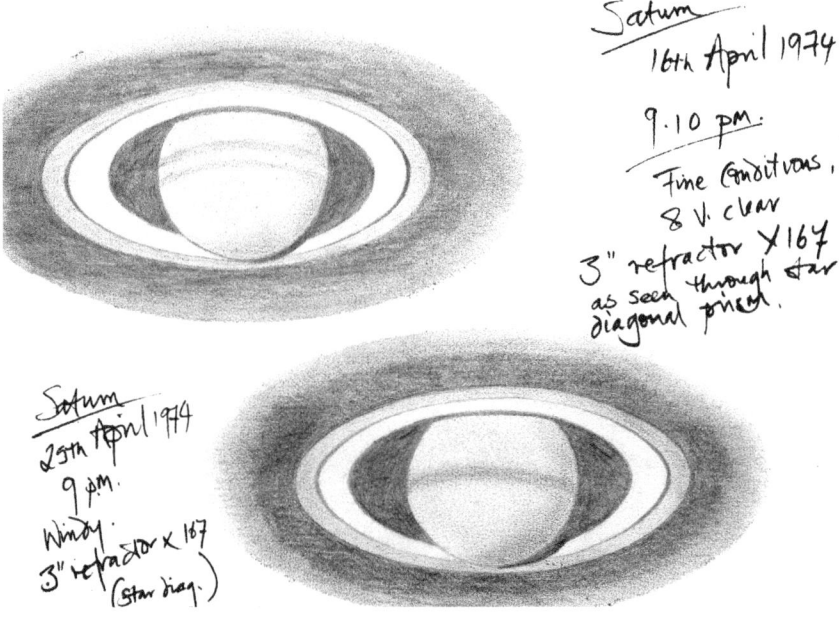

Fig. 10.3. Saturn 4/16 and 25/74. 3-in. refractor 167×. (AC)

it is a treasure chest for the amateur observer and draftsman. Unsurprisingly, Mars and Saturn, in that order, will take more time to draw.

Quick Project: Drawing the Planets in Color

Time Required: 30–60 min per Drawing

Always remember, less color is more. Because most planetary features are somewhat vague, do not succumb to drawing planetary features blatantly; these have little to do with reality. Apply color lightly and build up gradually, in multiple directions; use fingertips to blend together. A sharp eraser is a great asset in refining detail; print-type erasers are valuable for removing ink where it is too dark, and also where paper irregularities create stubborn spots and dots. Always stand back and take in the complete effect; try to reproduce the visual experience precisely. You do not have to be a great artist to do this to stunning effect. Remember, planetary drawing is more a work of patience and persistence than one of talent. Just do not settle for what does not satisfactorily represent the appearance of the subject at hand; tinker with your work until it does! (Fig. 10.4).

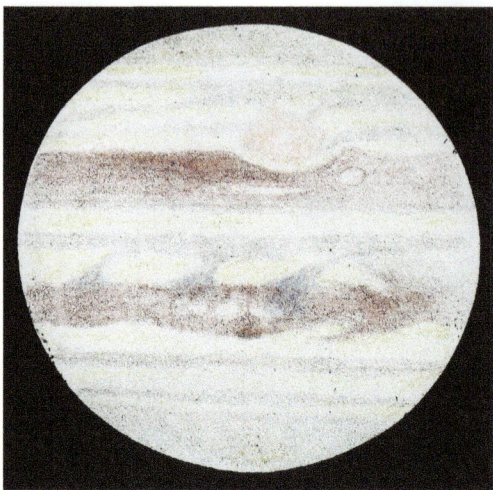

Fig. 10.4. Jupiter November 9, 1999. 18-in. reflector. (AC)

However effective you may feel drawing is, nevertheless, you should be aware of the differences likely to remain between the results and the live view, despite all efforts to create images that are as realistic and representative as possible. Overall, such reservations stem mainly from a realization that somehow they lack the sense of depth present in the eyepiece view, but perhaps you may not feel the need to strive for better results than these; they certainly fulfill the mission.

Mars presents its own unique set of problem in drawing. Luckily, its simple circular shape makes producing blanks simpler, although its pronounced phases require that you produce numerous variations with the correct phase portion eliminated. We will cover many of the challenges with this notoriously fickle subject in detail during upcoming sections in this book.

Meanwhile, Saturn has always been a wonderful, though highly demanding, subject to draw or image, and it will teach you more about astronomical "seeing" than will most celestial objects. At first blush it would seem that it should be easier because it is so spectacular and immediately striking. However, Saturn is so difficult to represent accurately by drawing that typically you may find yourself limiting full disc color drawings to just one or two per apparition, usually spending many hours on this one carefully executed representation. You will never have a greater appreciation of all that you are seeing in the eyepiece than when you attempt to put it on the page, nor will a successfully executed representation of this particular subject ever bring you more satisfaction. However, you probably do not have the time to do it!

Although much of the difficulty in basic drawing of this subject comes about by the need to represent the constantly changing curved sweep and proportions of the disc and rings correctly, some problems also arise because of the subtleties of the planet itself. The shape of the planet's disc itself is often portrayed as being too round, and the rings having too square "shoulders." For those who might feel guilty about reluctance to commit images to paper, luckily (or unluckily, however you may see it), Saturn is a relatively static object, in as much as really significant changes are slow in coming about. More likely, the variations such as we see will consist of pale spots showing up on the disc from time to time, differences in color and width of the already faint belts and zones (although do not expect anything of the order of variation and detail as those on Jupiter), other subtleties in the rings themselves (probably only because of seeing conditions), and even the so-called "spokes" (but probably only to the most keen eyed among us).

Drawing, for whatever it is worth, is at least one method we can always use to good effect, and leaves us all the more skilled as observers. But bear in mind that because many planetary draftsmen fail to draw what they see with anything vaguely approximating accuracy, admittedly, a good CCD image is *far* better than this, at least as far as reasonable representations are concerned. To draw well, one must be brutally honest and faithful in what is being committed to the page. If we just take a little trouble to acquire the necessary technique, it will be developed and guided by truly objective viewing and critical questioning of what we are actually seeing. With all of the fleeting, highly refined, and subtle details in the field of view, this is, of course, much harder than it sounds, but once you have a certain amount of "know how," drawing does allow you to proceed quickly without the multistep technical complexities of CCD or other hi-tech imaging. Because this in itself is never quite enough for perfect representations there will always be a dimension somehow missing even in the best results most people can achieve.

Nevertheless, the intrinsic weaknesses of drawings leave room for improvement, so it is not surprising that many planetary observers have succumbed to the lure of modern electronic imaging. Despite the frequent claims that a fine planetary CCD image will reveal more detail than the eye can see, this is only true in some respects. Indeed, considerable tangible detail may often be clearly resolved and contrasted in well-processed CCD images, and sometimes they are indeed more detailed, but yet different to the real-time view. This is also true sometimes in the simplest forms of CCD video imaging. Certainly, different exposure and processing techniques are capable of revealing detail and features that would otherwise go undetected, and many important planetary discoveries and ongoing research have depended on things other than the strictly visual approach. As amateurs, it is amazing that at the distance we are on Earth from any of the planets it is possible at all to glimpse *some* of these same refinements live through the eyepiece, except there is no question that they do appear in uniquely different ways.

Highly sensitive monochrome CCD video cameras can also be a good bet for quick and easy planetary imaging because of their very fast frame registration, usually 30/s. From the video stream, one may then select the best individual frames having the maximum clarity and resolution. However, it soon becomes obvious that the completeness of detail that the eye sees in the moving video itself consists of a composite of many frames. Few, if any, individual frames ever capture it all, and this is why it is such common practice to use composites of numerous images in later processing. Regretfully, for the ultimate results, once again this involves increasing amounts of time spent away from the telescope, which perhaps you do not have to begin with! However, some of the finest planetary images were produced by this method.

Full color video images may also be achieved in the same manner (as in standard CCD imaging) by using the tricolor filter technique, an even more laborious process, despite the potential it offers. Recently, even lowly web cams have become fashionable in astronomy because of the surprisingly good full color results that are possible from combining many frames taken with different color filters. However, one has to remember once again that the complexities of combining and processing many frames to produce a whole are something not readily undertaken by everyone. You can see we are already beginning to tread a path that you may find problematic. In our quest to produce the "ultimate" image, we may become pixel manipulators, hardly observers, and the demands on our time may make the situation impossible. Few amateurs, even with unlimited time at their disposal, seem able to be both observers and image makers.

Although it is difficult to quantify all that makes up the live view, it consists of that unique "crispness" and luminescence of the image, the full range of all of the subtle shades of color. It especially consists of all those vague, but present, refined details completely lost even in the best electronic planetary image. Hard to describe, even to draw? Very much so, but if you have spent any time at the eyepiece of a truly fine telescope you already know this, and you know that these characteristics are real, nonetheless. However, in defense of a certain few exceptionally skilled CCD enthusiasts, some truly amazing, even near lifelike images of the planets have been made. These talented imagers have sometimes been able to capture many, if not all, of the subtleties that we may have seen live but were somehow unable to define mentally for the purpose of drawing. In order to record their subjects so accurately they must first have spent time as visual observers. Presumably these same people are able to keep their observing roots firmly planted while embracing the best of today's imaging technology. In Miami, Florida, Dr. Donald Parker routinely makes superb CCD planetary images; these are possibly the best and most representative planetary camera images ever made by an Earthbound amateur. They may be seen in publications and on many websites, including the website of ALPO (Association of Lunar and Planetary Observers): http://alpo-astronomy.org, and perhaps most notably at the website: http://www.masil-astro-imaging.com/Don%20Parker.html.

Parker's work stands at the pinnacle of just what can be achieved in electronic imaging. However, without the skills, awareness, or lengthy experience that Dr. Parker possesses, equally good results would be very hard to come by. And certainly, what he has attained cannot be anything that someone pressed to spend even just a little time at the telescope could ever aspire to do. Yet, after all of this, it has to be said that the images *still* do not quite equal the unique impressions created by live viewing!

There are other considerations in planetary imaging, too. Part of the problem, despite any fine detail revealed, is that the discs usually appear slightly "fuzzy" in character, partly as a by-product of exposure length through a turbulent atmosphere, no matter how brief. The perfect moment of atmospheric stillness is either too short or unlikely to be caught by the user. Because much of the intricate detail often is still present in many amateur images, though, this cannot be the only reason. So, although CCD imaging made at an effective scale provides a considerable improvement over anything we had before (because of the shorter exposure times and its greater sensitivity relative to film), it is still partly subject to subtle differences compared with the live view, for whatever reason. Further processing often reveals remarkable amounts of otherwise unseen detail, but this also usually creates an increasingly unnatural appearance of the image. Thus, there appears to be no way to record the indescribable refinement of detail of the live telescopic view, even by the most painstaking imaging techniques. However, regardless of the level of draftsman's skill we may possess, our own attempts at drawing can still reveal certain attributes of planetary detail unlike that shown by any other method.

In trying to fit astronomy into our lives, being an observer may be all that we have time for. Nevertheless, the challenge remains on the horizon of recording even better what we have seen with our eyes in ways that do not get in the way of our limited time at the eyepiece. Can it be done without the heroic means normally required? The answer is "yes." Although the *perfect* answer may be impossible at present, there is a way to bring us a lot closer to the ideal.

Some Imaging Perspectives

In making countless full disc drawings, you will find that there are only so many images that are good enough to remain interesting to look over in the long term. However, the processes of extensive time spent drawing can produce valuable results, and you will certainly cultivate some usable drafting skills over the years. Isolating individual details or regions to draw, made over a portion of the observed planet's rotation, ultimately may make more sense. Making full "cylindrical projections" or full global maps of entire planetary discs can be a natural outgrowth of this approach. Certainly, it will prove to be immensely satisfying and informative, perhaps providing the icing on the cake to the process of drawing, with drafting skills developing accordingly.

Quick Project: Combining the Best of Video and Drawing

Time Required: 20–30 min

If you are resolutely resisting the pressure to become a digital imager, it is likely that you will still be looking for better ways to proceed.

By aiming a CCD video camera at planetary targets, you can record up to a minute or so of moving video. Hopefully always mindful of extracting the maximum performance from your equipment, you will push the image scale to the point where further increases provide no benefit. Also, while looking through the eyepiece, make rapid sketches and take detailed notes on the general appearance of the planet, including regions and zones of coloration. Later, extract a few good still frames from the video, and select the best of those as a reference along with your notes and memory as a reference to make the finished full disc drawings. (A simple printout of selected video frames can also provide a ready and accurate reference for the drawing.) Re-examine the moving video to check for subtleties missing on the individual frames or printout.

However, differences are likely still to remain when comparing drawings to single raw video frames. This is true whether the drawings are made in color or in black and white. Final completed full color drawings can still sometimes take an hour or more to produce - again something not ideal for us! Additionally, the chief weaknesses of drawing remain, even if now less pronounced – the inevitable subjectivity, difficulty of actually mentally resolving all of what is seen, one's drawing skills, exact placement of features and relative intensities, and so on.

A New Solution!

You are, of course, familiar with movie colorization. Keep reading; however atrocious this may sound at first blush; astronomically speaking, we are going to do something related to that process, at least in essence:

Quick Project: Combining Video Frames and Drawing - Jupiter and Mars

Time Required: Each Finished Image 20–30 min Total

The first step is to take a few good notes at the eyepiece. Then record a video stream of the planet at the most advantageous scale, which you can continue to refer to throughout the following process. Extract the single best monochrome video frame from the moving video clip and print it at the highest quality and at a suitable scale for drawing. The printed monochrome image will serve as the foundation of the drawing itself, very similar to the method of constructing drawings from scratch, where the features on the planetary disc were built up initially in degrees of gray. But now the gray shadings of the video image do basically the same thing! Just make sure that image extraction or original exposure does not saturate the page with too much dark contrast, and adjust it accordingly.

It is a matter of coincidence and great convenience that the standard type of paper used in most computer printers is of a highly refined, smooth, and even texture. This is well suited to taking colored pencil evenly, and enabling the necessary refinement of detail. You should always search out the whitest paper available, but no special or expensive varieties are needed. In fact, you will find just the basic paper to be quite satisfactory. Additionally, with printed video frames, you may welcome the relative changes in size of a planet as it approaches and recedes, and sometimes favor allowing these variations in the printed image size as part of the record of the apparition – something you are unlikely to do when drawing from scratch on blanks. You should decide how you wish to proceed and adjust the images size to make them all conform, if you wish.

It is an easy matter to refine and add to the monochrome detail, filling in everything, including color, directly on top of this printed image, just as one would a conventional drawing. With images made on an ink jet printer, one can also manipulate the printed image itself with erasers and black and gray pencil to shade and sharpen the features; the ink medium is readily flexible. This image manipulation may include that of the disc outline itself (usually too fuzzy in the raw image), and done in a similar manner to a conventional drawing. The background sky will be printed jet black, which hopefully will be a close match for the black pencil you use.

With video or other electronic imaging you might benefit from the use of certain colored filters to bring out recordable features and detail. Obviously, CCD imaging requires the use of specific filters to produce color images, but we refer now to the use of black-and-white CCD video cameras, where you will find that a red filter, for example, gives wonderfully contrasted views of Mars on the monitor. Wherever, in theory at least, a certain filter will emphasize a particular aspect visually, you might try the same approach of choosing appropriate filters for your camera. Utilizing them in this manner is sometimes far more valid and significant than for visual applications at the eyepiece, and may indeed reveal more strikingly whatever is present in the image, otherwise seen less easily on the monitor or not at all. A monitor actually provides some of the same benefits as a binocular viewer, in as much as it grants the same indefinable quality of being able to look at an image with both eyes simultaneously. Overall, the main benefits of using filters with video are likely increased contrast of features against the planet disc, along with the resolution of some details that might otherwise be missed. However, do not rush out and buy a wide range of filters until you have had a little experience in determining what general colors, or depths of color, produce worthwhile results for you. Only a few applications in video – especially on Mars – of very specific colors and depths seem to do any good at all.

On video images, you will notice a falloff in illumination all around the disc (and especially exaggerated on the side where a phase is present, no matter how slight the phase may be). This seems common to all photographic, CCD, or video images. It will be much more dramatic than is visible in the eyepiece, since sharp edge definition is a striking feature of any live view, even when a phase effect is present. The use of filters with your camera will likely make such differences in the appearance of the disc's outline even more striking. In your final image, although it is possible to reduce this limb darkening wherever it is extreme, the falloff in limb brightness is always likely to be more pronounced than you detected at the time of the live observation. Indeed, you may not have been aware of any drop-off at all in illumination on some subjects! So, define the perimeter of the disc more by carefully outlining the blurred disc on the page with the black pencil. The advantage of leaving some of the effect present, however, is the added impression in your final image of a dimensional globe structure, rather than just a flat disc.

Sometimes, when compared with the live view, other differences will be apparent as well. You will find that, depending on whether or not you used a filter in conjunction with your camera, many areas or surface features may register quite differently, and do not represent the live appearance with reasonable accuracy. For example, in the case of Mars, limb hazes and clouds will appear dark instead of light on the video image, particularly when captured through a filter. Erasers and white pencil can do wonders with all of this. The main challenge is to establish the complete background features in various intensities of gray to correspond reasonably closely to their intensities in the live view. This will involve sometimes lightening whatever features are printed, as well as adding, completing, sharpening, and defining what the eye perceived live. You will always need to adjust. To sharpen the more defined planetary features, try gray color pencil.

Once you have taken care of the basic gray image foundation, add those trace colors as you saw on top of the printed image, in whatever degrees are necessary. The slightest suggestion is enough, and, as always, cannot be stressed too much. When this step is complete, add the dominant disc color in an appropriate amount over the entire planet disc, including those original gray features! Surprising as it sounds, this is necessary in order to create the full natural coloration, which tends to wash over the whole, giving the specific hue we are accustomed to associating with each planet. In another surprise, most of the other colors will tend to appear more realistic by this process, rather than obliterated. Interestingly, the last fine adjustments of shading may be attained quite easily with a simple lead pencil. Just lightly dot any small area needing subtle darkening, and blend with fingertip; this method may be used to complete the fullness or evenness of virtually any color.

By at least partly correcting that certain indeterminate fuzzy quality present in all imaging processes and including many of the subtleties that you saw in the eyepiece view as possible, you will begin to approach the way the planet came across in the eyepiece at the time. Overall, it is a simple process to do all of this in exactly the same way as one would a regular drawing, except that now you have a visually deeper foundation to work upon and no longer need to begin by correctly placing the features themselves. Finally, especially while you are finalizing your composite images, it is instructive to examine the moving footage from which the single frame has been selected. Although the "footage" is rolling, all of the somewhat vague attributes will be far more defined and apparent to you. Any last minute adjustments and additions may be made at this stage.

When initially viewing the planets by live video, try varying the brightness, contrast, and shutter speeds to produce darker images; all manner of detail will be revealed that may be studied to good effect at the time of observing. However, you should remember that most of the darker video stream will not provide usable material for still imaging, let alone printing. Such uses of the camera are best kept for study or sketching at the moment itself. Regrettably, once you increase the brightness level, some of this subtle detail will be lost in the glare, so it is always important to spend a little time noting everything present on the monitor under varying illuminations. You may, of course, record portions of such darker video stream for later study, as long as you bear in mind that this material will not provide usable printed images, no matter how you try to manipulate it.

Above, in Fig. 10.5a–c, is the same view of the "red planet" to illustrate the various methods of imaging as just described, culminating in the final "colorized" approach

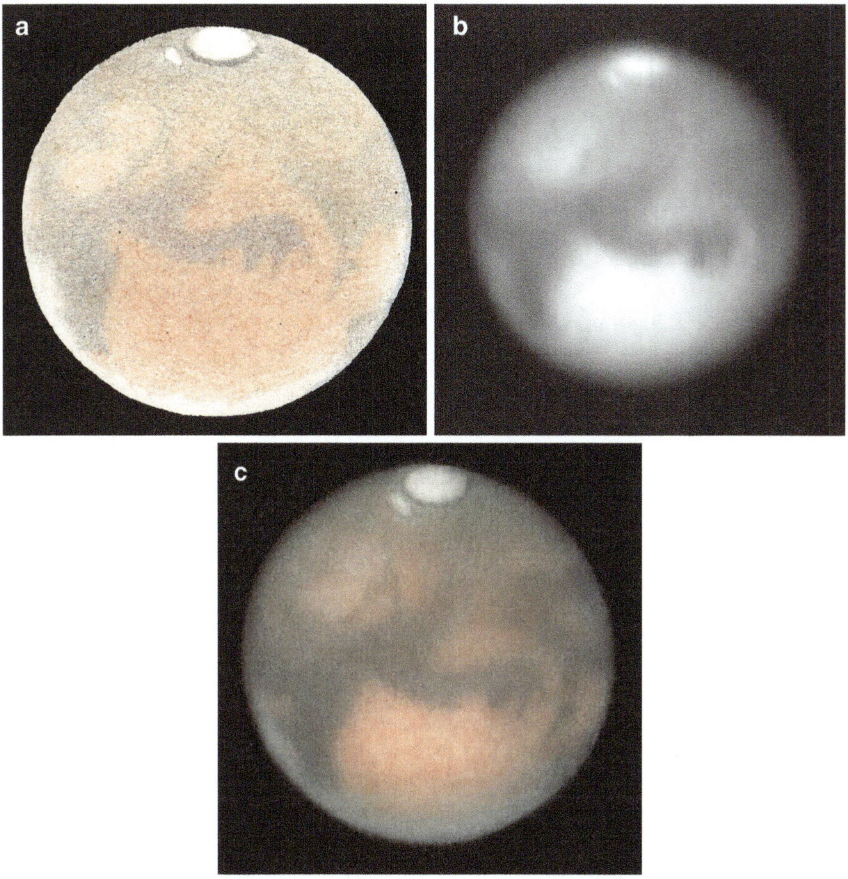

Fig. 10.5. Mars: September 6, 2003. (**a**) Color drawing, (**b**) Video frame, and (**c**) Combination video frame and drawing. (AC)

in (c). You will immediately be able to see the progression that these methods have followed toward combined video image and drawing. They provide highly realistic and easily attained results.

The final view, Fig. 10.5c, presents a remarkable likeness of the Red Planet, as good as many better CCD images, in fact. Imaging by this method is easy and fast. Though not quite as detailed as the very best processed CCD images, images produced by this process come remarkably close and seem to have more of that hard to define "living presence." Part of this is because of the more dimensional effect the video image brings to the picture in the first place. Even more striking is that certain luminescence present in the live view, which now often seems present in the finished image as well.

Although full color images make an enormous difference for Mars, Jupiter, and Saturn, you may find that with most other subjects, such as Venus and Mercury,

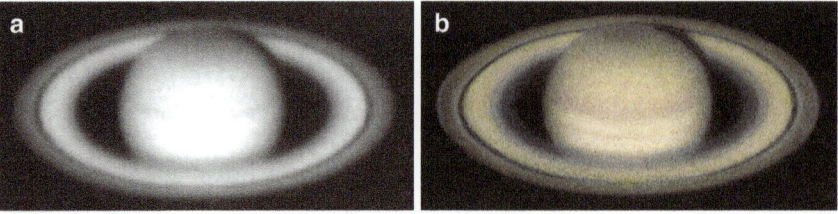

Fig 10.6. Saturn: January 31, 2003. (**a**) Video frame and (**b**) Combined video frame and drawing. (AC)

color is of much less consequence, if it has any value at all. You may count yourself lucky if you can see any detail on these subjects, let alone anything other than one predominant color! Visible markings on these subjects are few from planet Earth and appear vague at that, and ultimately, imaging these less visually striking planets in color is a decision that is entirely up to you. Similarly, combining the earlier technique of combining video frames with drawing may well be overkill, when a simple drawing at the eyepiece will do the job perfectly, quickly, and easily; it should not be difficult to draw them very rapidly and accurately from scratch, or just use simple, lead pencil retouched video frames.

Time Required: 60 min

Using the "colorization" method, Saturn becomes far simpler and easier to represent on the page. Although the results usually lack the resolutions of many fine regions in the rings sometimes possible to obtain with advanced CCD imaging and post-processing, you may be quite happy with the results. Indeed, they are quite representative of the live view and fairly easy to obtain. The best thing about this method is not having to set the proportions accurately of the rings and planet, something which will always slow you down.

Quick Project: Combining Video Frames and Drawing of Saturn

Having obtained some good video clips (wait for a still night; the difference is incomparable with Saturn), print an image at a decent scale from a carefully selected video frame. With a sharp black pencil, define all the boundaries, shadows, and ring divisions (typically just Cassini's) that you were aware of in the live view (be sure to take notes). You will probably need to lighten the region of the Crepe Ring just a little, along with the bright inner edge of the "A" ring, in order to make them show properly and be true to the live appearance. So add a little pale blue and white to the Crepe Ring. Within the rings themselves, the yellow seems brightest at the outer edge of the "B" ring, and a pinkish color predominates at the inner edge; there seem to be definite separate components to the "B" ring that are important to define. These appear more as steps in brightness and colors and should not be confused with divisions, such as Cassini's.

The overall color of the planet can be reasonably well represented in varying degrees by a deep yellow with a little salmon pink added on top of that. Equatorial belts often seem to exhibit a crimson hue, so a trace of a dark shade is usually needed here.

Figure 10.6 shows an example of a basic video frame, taken in good viewing conditions, together with the *colorized* and refined version of the same image for comparison:

Stand back and imagine that you are looking at this colored image through the eyepiece of your telescope, slightly defocusing your eyes as you do so. It really does simulate the ringed planet's appearance in the field of view, does it not? Other than the ease of making these images, this is the other true strength of this imaging system.

In striking contrast to the other great planets, numerous fine CCD (and other electronically made) images of Saturn do exist, and many amateur enthusiasts have been quite successful in their efforts. The great planet seems to come across significantly better in such electronic imaging than does Jupiter or Mars. Maybe part of it is because the detail we are able to make out from Earth is less complex than that of some of its neighbors. It is especially common for Saturn to be portrayed fairly realistically in CCD images, which often look surprisingly akin to the object long familiar in the eyepiece. Nevertheless, its unique golden brilliance, as well as the crisp yet difficult-to-determine view, still remains elusive, even with the best of CCD techniques. That very coloration seems to be problematic to represent realistically, no matter how we go about it, and it is difficult to recall ever seeing an image that accurately portrays how the eye sees the special gilded hue of the whole. In virtually all electronic color images it seems most absent in the rings themselves, which usually record as a relatively bland bluish white.

Your eye's connection to the mind remains still the best and fastest planetary imager of all. Your own impression will still far exceed any recorded images you may make or see by others, and you will have spent your available time as an observer and not a technician.

CHAPTER ELEVEN

Spectacles in Our Neighborhood

Traditionally the hard core of the amateur astronomer's universe, three spectacular telescopic subjects, Mars, Jupiter, and Saturn, should be no less important to us today than they ever were, despite the decline in opportunities for the amateur to be on the cutting edge of most modern planetary research. These three great destinations represent some of our best opportunities to have nearly unlimited enjoyment in the sky. Only the Moon gives us a better opportunity to understand the true nature of another world's surface. The other great upside is that these destinations are so accessible that they allow us to view them on our terms most of the time they are in the sky; it is not necessary to dedicate unlimited hours to see something worthwhile. Repeated viewing of these tried and true subjects never becomes old, since the three great planets may be observed undergoing constant change, and the opportunity for witnessing such phenomena within earthbound timetables is rare enough in the universe. It is hardly surprising, therefore, that these "big three" planets often become many a newcomer's primary fascination. Their hypnotic lure even keeps many experienced observers transfixed by them for the long term as well, occasionally to the exclusion of all other types of observation!

Although some of the other planets do indeed provide some degree of color in the eyepiece, no one could possibly pretend that they exhibit much of vividness in dramatic and varied displays, let alone detail. Thus, even within the solar system we cannot take color for granted; deep space is infinitely more challenging, so take full advantage of this opportunity waiting right in our cosmic backyard! The best part is that no special imaging equipment is needed to experience the most dramatic kinds of scenes, vivid as anything we could imagine. Fortunately, too, for us, the big "three" will usually put on a spectacular show more readily than most objects in deep space. With them at least we can usually see *something* impressive in the field of view, no less from the heart of any brightly lit city. It is worth noting also that these three worlds offer the reasonably

A. Cooke, *Make Time for the Stars: Fitting Astronomy into Your Busy Life*,
DOI: 10.1007/978-0-387-89341-9_11, © Springer Science+Business Media, LLC 2009

well-equipped amateur the opportunity to actually see many of the finely imaged details seen on the NASA web sites listed in Chap. 16. The NASA web sites are a treasure trove of visuals and information where the amateur's telescope is unable to serve us well. It is significant how much this resource adds to whatever we are able to make out at the eyepiece.

Take the time to view and study these three great planets with your own eyes, and above all, avoid the rush into today's imaging craze; you only rob yourself of something quite wonderful and far more personal – certainly more in keeping with busy lifestyles. It was the planets that taught many observers to extract the maximum with their eyes from whatever equipment they had or still have. This, then, is the delicate art of astronomical "seeing," where the eye learns to adapt to that special brand of skills needed to discern fine and fleeting detail through the eyepiece. Since we can acquire these viewing techniques with just a little patience, eventually, they will enable us to see considerable detail easily, better than our earlier observing sessions. Those skills, once acquired, will also reward us forever in all the other areas of observing.

Hopefully, some of these best, most colorful, and brightest objects in the universe will provide you lifelong enjoyment, as your backyard spaceship carries you toward them in the night sky. We also have the real prospect of knowing their surfaces close up, through the Internet, via the great age of planetary space exploration that man has now undertaken; the solar system, overlooked for so long by the professional astronomical community, has now resumed its position center stage! This can only provide better perspectives for observing at the telescope.

Filters, Again!

The use of color filters is probably more connected with observing the planets than with anything else. Filters, occasionally useful though they may be, are far less of a factor in most live visual applications than you may have been led to believe. The times when you will be able to see *extraordinary* amounts of extraplanetary detail with them are few and far between, if ever. Thus, it is not likely because of a lack of filters that you may not be seeing incredible things. However, there *are* times when their use may indeed be marginally beneficial, as long as you know what to expect of them. None of this is meant to stop you from conducting your own experiments with them, of course.

Quick Project: Evaluating Views of Mars, Jupiter, or Saturn with Color Filters

Time Required: 15 min

While viewing Mars, Jupiter, or Saturn, try using different color filters. Mars is one of the few subjects with which you may be rewarded by using various color filters. They may help you to better make out very faint markings. On certain occasions, you may able to discern very subtle dark shadings (with red and orange filters) on the Red Planet, which initially might otherwise have escaped your gaze. Limb hazes are undoubtedly more contrasted when using a light blue or violet filter, although one would have to be quite unobservant to fail to see them at all in the first place! A blue filter may enhance Jupiter and Saturn; the belts are certainly more prominent this way. Most other color filters provide very limited results, and you may even conclude that they impair the view!

Meanwhile, do not forget that filters do take on new importance with imaging, where you will find that a red filter gives wonderfully contrasted views of Mars on the monitor, with black-and-white CCD video imaging.

You will find in strictly visual applications that the differences are extremely subtle. When initially learning to discern detail, such awareness may be valuable at the time, but, for the most part, once you know how to look for such subtleties, you will find that you can readily detect them without the aid of a filter at all – almost every time. However, you be the judge!

Filters may be most valuable when there is a question in your mind as to what exactly it is you are seeing. Sometimes with the right one, certain details will appear a little darker or be more clearly visible, especially when you are not quite sure of what you suspect is present in the image. Such filters will also aid in confirming the finest dark shadowy detail, which are often extensions of more prominent markings, so that you can comfortably switch back to unfiltered viewing.

However, be warned: Do not expect to see nearly the level of improvement you may have heard is likely! Usually, it is best to view the planets "au naturale," *in good viewing circumstances.*

Mars

Of the three great planets (Mars, Jupiter, and Saturn), Mars, the "Bringer of War" of classical astrology, remains the one that still conjures up more mental imagery in most peoples' minds than any other single object in space. Even its predominant red color seems appropriately warlike. Today, with ever-increasing knowledge gleaned from spectacular space probes, even the layman has an idea about the true nature of Mars's surface. Of course, it is nothing like the planet we may have imagined years ago, but in other ways, it has turned out to be no less wonderful. A visit to NASA's website is always highly illuminating, and the complete record of their space missions is available to explore. It is hard to get over the drama of seeing some of the first images from the rover "Opportunity," showing detailed rock strata, reminiscent of some magnificent prehistoric dinosaur bones.

Actually, in a perspective far away from this kind of good science, a case can be made that the astonishing rise in the popularization of astronomy in general during the twentieth century was actually fueled by the many incorrect conclusions that were once made about this particular world! Apparently even bad science has its place. Certainly, all of the frenzy helped to fuel the quest for the average layman personally to witness this alien place via backyard telescopes, which only led to other destinations in space. Maybe today's newfound and spectacular visions of Mars' surface will continue to keep the red planet as the catalyst that inspires many new devotees to astronomy long into the future. A wonderful book, *A Traveler's Guide to Mars* (Workman Publishing, New York, 2003) by William K. Hartmann is easily one of the best of its kind and should further connect new generations to this alien world. Human beings are destined to walk its rocky surface during the first half of this century.

It must be said that one controversial figure in particular – Percival Lowell – will forever be associated with this small corner of the solar system. Because he was responsible for placing Mars so large in the public's imagination, it would serve us to look at his contribution. Concerning the "bad science" he promoted, we should cut him some slack. With all that we have now come to know about Mars, it is quite understandable that certain observers with a particular inborn connection of eyesight and mind came to superimpose the so-called *canals* across the surface. Lowell certainly was not the first to do so. In fact, it turns out that some of the features actually *did* form a basis in reality for some of the *canals*, such as the canyons of Valles Marineris and its various tributaries around the Solis Lacus. Lowell's eyesight was, in fact, pretty good.

Lowell had been ridiculed for so long that is only recently that anyone would actually dare to raise the suggestion that some of the features he described and drew might actually have existed during his day! Although we know now what they are, we can see some of these plainly enough for ourselves with a decent telescope at a favorable opposition. Certainly, it was not incorrect, at very least, to describe some of these features as "channels," and so it turns out that some earlier observers were not so very wrong about everything they described. The conclusions they drew are another thing entirely! Other descriptions of the time also often centered on real features, only proven to us in recent times. For example, linear markings within the great basin Hellas, as well as around the entire region, in fact, do appear at different times. You can read outright denials of their existence made not so long ago by well-known observers. (I saw these criss-crossing lines with my own eyes during the highly favorable opposition of 2003, and recorded them *easily* at that time with a video camera; this should settle any further argument! Just examine the video frame (a) in Fig. 10.5 in Chap. 10. Look carefully; the lines are there!)

Then we have the hypothesis of the apparent appearance and retreat of the supposed seasonal "vegetation" of the old observers, which actually turned out to be only the seasonal effect of drifting sands born by annual winds. However, all of the earlier conclusions about the advance and retreat of these phenomena (supposed vegetation) were quite reasonable, even scientifically justifiable, given what was known at the time.

There are many other examples of linear "features," too, including the so-called *oases*, and the spontaneous appearances of "double canals," some of which have been deduced recently to have been caused by drifting sand formations. The ongoing changes to the surface certainly allow for the possibility of once described real features no longer remaining in existence; this includes actually observed and recorded long-term changes to the albedo features themselves over the years, and which we often tend to ignore in our quest to ridicule the earlier observers. Could it be that some of what we assumed to be imagined details, and at one time boldly described by these individuals, were really there after all? Just look at old photographic images of the Syrtis Major, for example, and this possibility will be more readily apparent; its shape has changed significantly. Look at the old pencil drawings from 1978, too.

(See Fig. 10.2, Chap. 10 and compare it with more recent images in this chapter.) These changes in shape are easily explainable because most of the so-called albedo features are not really features at all but different types of soil, sand, and dust, merely colors of mineral deposits. The Martian winds, slight as their actual force is, provide all of the necessary energy to slowly transform these landmarks over time. Indeed, some features seem to change noticeably during remarkably short intervals, often not more than the time between two consecutive oppositions.

A little research will reveal that Lowell, that most fanatical of early Martian observers, was, in fact, a highly educated man and a distinguished mathematician in his own right, with important books and writings to his credit. It was not so unusual in his day for an independently wealthy and educated person to follow an interest passionately; certainly, he could afford to do this. He took things to extremes, as only well-off educated nineteenth century aristocrats could, and sank his time, money, and ultimately his reputation into the study of the universe as he saw it. Chuckle though you may about his Martian civilization conclusions, Lowell at least dared to use his mind and energies in search of answers. The fact that certain great astronomical mysteries of the day were what motivated him merely speaks to what led him to dedicate the better part of his life and energies to them. Lowell certainly had the magical spark of adventurer, all too often missing in astronomy's "new school." This one "amateur" had much to do with the popularization of astronomy, and for this we owe him a great debt. Be honest; do not you see something of yourself in that well known and inspiring, even sentimentally touching, photograph of him? Who among astronomy's older devotees can ever forget this famous image of Lowell, perched inquisitively on the observing platform of his great Alvin Clark 24-in. (60 cm) refractor, peering up at Mars – *his* Mars?

So, it now seems that Lowell might not have been such a crackpot after all; others have come to the same realization, especially since some of the features he described and drew were, in many instances, actually real, at least in some sense, or could well have existed at that time. His conclusions as to what they were may well have been incorrect but were more logical than many of his casual detractors had wanted to concede. It is easy to be one of these detractors now that we have discovered so much, especially since Lowell was an easy mark with his colorful visions of Martians and their dying water supplies and cities. So while he was very wrong about many of his conclusions (most of them do seem decidedly eccentric, even quaint, or just plain "off the wall" today), given the amount of information available at that time, and the time in history itself, he really was not any more wrong about many of them than have been many other highly respected figures in the past about other things. The real problem was his stubborn refusal to budge on his conclusions once other studies began to refute much of what he had written and lectured about over the years; this one fatal flaw will always haunt his reputation and set him apart from the professional scientist. However, it certainly does seem unfair to continue to cast him dismissively as a mere "crackpot," since there was quite a lot more to him than being just a wayward dreamer.

Along the way, Lowell was also responsible for attempting some of the daunting mathematical computations in predicting the existence of a planet beyond Neptune. For some years before his time, it was apparent that there were some unaccounted perturbations in the orbits of the known planets. Lowell's calculations indicated the existence of another planet ("planet X") outside the orbit of Neptune. Even though

Pluto was not quite the planet he had anticipated (its mass is insufficient to cause such disturbances, especially clear to the public now that it has been demoted from being a planet), nor in quite the right place, Lowell alone was behind the great push to discover what was then to be known as the ninth member of the planetary system. As it turned out, Pluto was not to be discovered until after his death (by Clyde Tombaugh in 1931), but fittingly, the event happened at Lowell's Observatory itself. Almost immediately, it became the subject of a perennial controversy that raged for generations over its rights to planetary status within the solar system. This has continued to this day, and despite its recent fall from grace to an object of lesser standing (a dwarf planet, or "planetoid"), the debate may not yet be finally over. Astronomers are once again airing the subject! Meanwhile, Lowell's beloved observatory in Flagstaff, Arizona, has thrived, grown, and lived on to make many major contributions to astronomy, including the monumental discovery by V. M. Slipher of "red shift," the method by which we measure vast galactic distances in the universe, and one of the most important contributions to all of modern astrophysics.

Many potential astronomical recruits will surely want to see the famed Red Planet with their own eyes; indeed, it may be the greatest of all incentives to enter astronomy. With its special brand of magic dangled in front of us, the new recruits may not yet suspect that amateur planetary astronomy, and worse, *all* of amateur astronomy, is becoming anything but a visual activity. However, this is the very part of astronomy that remains so valid for people who have little time to indulge their interest. If they listen to the "experts," they too may find themselves talked right out of their dreams. Instinctively, there is still in most of us just a little bit of the same pseudo-romantic notion that inspired Lowell himself. For those who resist these pressures, Mars alone provides sufficient justification for maintaining an interest in astronomy. Since great optics may be had these days for a fraction of the costs of yesterday, live spectacular views of the Red Planet are more available than ever, at least when we place optical performance foremost.

For the study of Mars, especially, it is hard to overemphasize the importance of optical and mechanical quality. Do not confuse these things with complexity and automation! This particular planet is usually a highly elusive object, typically being small in the field of view, low in the sky, dazzlingly bright, and so detailed that it is sometimes difficult to resolve clearly in one's mind exactly what one is seeing. *Quality* aperture, and not just aperture itself (preferably of the most diffraction-free optical configurations possible), as well as rigid and smooth mounting design, remain our best bet to unlock Mars's secrets. On some occasions when you will be disappointed with Martian observations, it will be with equipment purported to be of high quality, but in fact, not. Some such telescopes are the property of surprisingly prominent amateur organizations, with apertures as large as 20 in. (50 cm) and more, sometimes installed in magnificent observatory housings. What may be missing are the most important common denominators of high-quality fundamentals: optics and mechanical soundness.

Happily for us, despite the difficulties of observing Mars easily, it is fortunate indeed that, on occasion, it can still live up to something approaching our level of anticipation, and frequently exceeds it, as long as we have the above equipment fundamentals, and particularly, well-developed observing skills. Perhaps a certain type of patience should be added as a final qualification, because the temperamental planet only approaches Earth briefly, and usually for not more than a relatively few weeks once every 26 months. Compounding matters, it is often placed unfavorably in the sky for most apparitions, and is notoriously prone to hide under unannounced dust storms, which can encircle the globe for weeks on end, masking everything that lies beneath. And guess what? These dust storms usually coincide with the closest, most favorable apparitions, which happen to be periods of maximum solar heating and turbulence in Mars's tenuous but otherwise more tranquil atmosphere.

The subject of magnification also needs a little comment here, since there really is no one absolute maximum power for *all* subjects, a fact that is often glossed over or boiled down to a "one size fits all" mentality. Although it is true that the much-touted maximum of 300× to 350× is a good general guideline regardless of aperture (because of Earth's atmospheric turbulence), there are so many exceptions to this that we always need to be open to the idea of trying powers outside this range. This is as true for solar system subjects as it is for deep space. Exactly why the variations of maximum magnification for different subjects occur is not always easy to explain, but sometimes it may be that they are affected by the concentrations of illumination of any given object, or even the specific frequencies of light reaching us. In Mars's case, because it is a small object, even at the best of times, there is obviously a practical need to seek the highest power it will take. Equally significant, Mars is one of those objects that, luckily for us and also perhaps inexplicably, will often withstand considerable and even apparently inordinate amounts of magnification. You may stretch the normal limits on this subject when it is viewed in optimal circumstances. However, at a really close opposition, such as the one we experienced in 2003, powers greater than 350× were never necessary, or even desirable; sometimes substantially less power was perfectly adequate to show everything the telescope was capable of revealing. In contrast to this, when Mars presents a significantly smaller disc, powers up to 450× and even up to 600× can be just fine on decent nights, and the highest possible power produces some of the best results. Because Mars will often stand so much more power than many other planets, you should not hesitate to try every such possibility in your eyepiece arsenal. Moderate apertures, especially of longer focal ratios, may surprise you in the powers they will allow.

The celebrity of Mars remains so much in the public eye, even today, that novices are more likely to be initially disappointed with what they actually see through the eyepiece. Because the "Red Planet" is so infrequently well placed in the sky, along with its usual diminutive disc, the challenges to viewing it well are only fully appreciated by those who have tackled it many times for themselves. However, with a little experience, Mars proves to be completely captivating, and it is amazing just how very detailed the disc ultimately can appear in the telescope's field of view. Therefore, if you only have a half hour – or even less – to spend with your telescope, return to Mars as often as you can; the rewards will compound themselves. It is possible that the Red Planet will remain the greatest solar system delight you have. Sometimes, the arrays of subtleties are so great that even Hubble Space Telescope images may come to mind! It seems totally impossible to grasp all of the visible fine details sufficiently to record them – even by the most refined drawings, and most especially, by any form of imaging, even that of the mind! Unlike CCD images and other views, the detail is quite fleeting, but it is at the same time much sharper and subtler, usually quite difficult to consciously dissect and describe. However, you know you can see it! Often it is so finely resolved, in fact, that you will likely conclude that live views of any of the planets are really nothing like the pictures with which you may have become familiar. Although the view is more wonderful and complex, it is somehow both less and more contrasted, clearer and more vague, all at once!

In this author's own drawings, despite considerable efforts, differences to such live views still remain, even though they are hard to describe. Thus, since your eye's connection to the mind is still the best planetary imager of all, your own impression through the eyepiece will still far exceed any image you have ever seen. On first glance, all you may see is its unique and overwhelming salmon/pink color. Look more closely; first a polar cap will become apparent; then, bit-by-bit, darker shadings, on first glance seemingly of a greenish/gray hue, will begin to jump out, followed by variations within the shaded areas themselves. Detail within detail would be an appropriate way to describe what you see. Try different magnifications as you make out increasing amounts of the remarkably complex fine detail. Subtle colorations gradually seem to make themselves known, though be cautioned that, although what you are seeing is real, many of the brilliant hues are enhanced – even falsely suggested – by the effects of Earth's atmosphere. Coupled with this is an effect known as *simultaneous contrast*, a phenomenon well known to experienced observers and brought about by the stark difference between object and black surrounding space. Just be sure to immerse your eye long enough in the view for it to relax and the detail to become clear. In this way, contrast and other subtleties of detail seem to become quite radically enhanced the longer we stare.

The apparition of 2003 was historic, in that Mars came closer to us than it has done in recorded history. However, it was not generally pointed out to the uninitiated that it actually approaches *almost* as closely every 15 years or so, the actual difference between those apparitions and this particular one being more of a technicality than anything else! How many astronomically unsophisticated souls set up telescopes at *exactly* the moment of opposition, thinking they would otherwise miss the chance to see the Red Planet at this supposedly once-in-a-lifetime approach? However, there was a key difference between this and many other favorable oppositions: the absence of major dust storms, all too typical in such circumstances, and truly a noteworthy nonoccurrence! With the benefit of frequent good viewing conditions in southern California, what a viewing opportunity it was! The wealth of detail was unbelievable.

In fact, all of the full disc images in this volume are from the 2003 apparition and were made in conditions extremely conducive to viewing. You will soon know not to waste your time trying to make planetary images of any kind unless the atmospheric conditions are right, and luckily the summer of 2003 obliged generously.

With the results of the "colorization" imaging method, next is a tour of the Martian surface, as the planet rotated around its poles during its grand visit in 2003. To easily produce such a grand tour yourself just follow the quick projects of the last chapter, only now with a systematic collection of images over an extended time. Each image here was made as a beginning and an end in itself, with the results for this compendium readily attainable.

It is also important to point out that because Mars's rotation is slightly slower than that of the Earth's, the sequence appears to be shown with the planet turning counterclockwise! And the pictures following were not taken in a simple sequence of just one chronological rotation! It proved necessary to collect the images over two complete rotations during the prime viewing weeks. This was essential to capture suitable points of the rotation (because of varying viewing conditions during the apparition at this location), as well as the quality of "seeing" the conditions provided. Additionally, in the layout of illustrations of Fig. 11.1a–g, originally this resulted in an uneven progression of disc size and polar cap retreat, which was adjusted here to show the Martian disc size as constant.

The realism of these images seems amazing, and considering how quickly they were produced makes it an ideal method for our needs. But the best part is that most of what is shown on typical, good CCD images is easy to see in these much more easily obtained examples. However, most of these composite images still retain something of that special quality that sets drawings apart from the other modern forms of imaging. Instructively, it is easy to see how similar lines and streaks, similar to these here, might have given rise to "canal" sightings of antiquity; they are quite plain to see (look especially at Hellas and nearby). Additionally, subtle colorations, some possibly due to the effects of simultaneous contrast and Earth's atmospheric

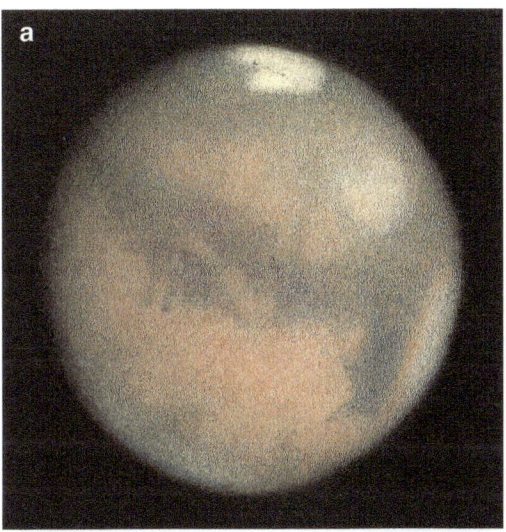

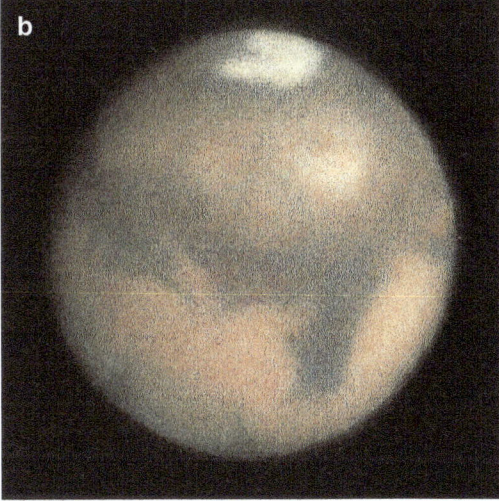

Fig. 11.1. (a–g) Video frames with color drawing. One rotation of Mars, from the 2003 apparition. (a) August 9, 2003 8:30 UT Seeing II. Amazing clarity of detail throughout entire Tyrrhenum and Cimmerium regions, including well-known northern streaks jutting from Cimmerium. Hesparia easily revealed. Hellas quite pale; Syrtis *blue cloud* effect clear. Striking dark band around fracturing northern polar cap. (b) August 7, 2003 8:15 UT Seeing III. Much south to north linear detail showing in southern hemisphere. Striking Syrtis *blue cloud*; Syrtis Minor clear, though of lighter coloration, almost mauve. Mares Tyrrhenum and Cimmerium well revealed. Dark banding around southern polar cap as well as fissure markedly visible.

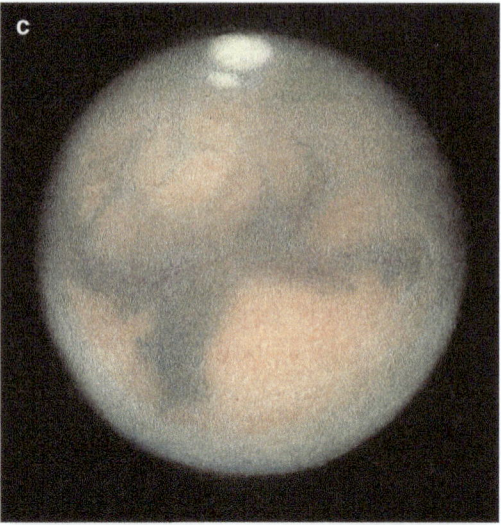

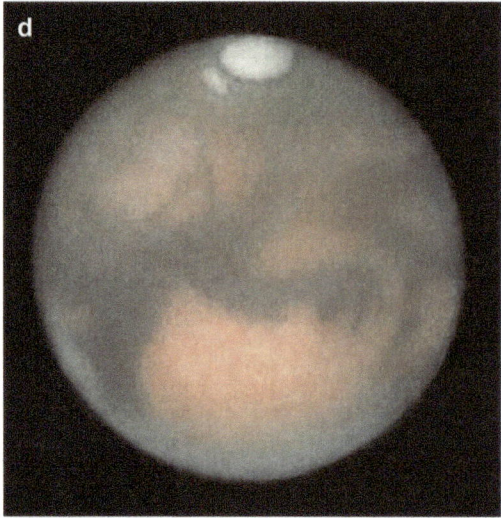

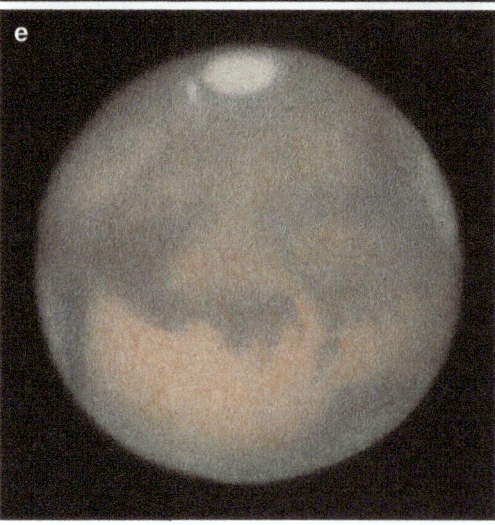

Fig.11.1 (continued) **(c)** September 8, 2003 6:30 UT Seeing II. Extensive southern hemisphere detail. Hellas shows central dark spot and streaks to the edge, not unlike a pale Solis Lacus. **(d)** September 5, 2003 7 UT Seeing II. Face-on view of Sinus Sabaeus, with much detail throughout entire disc. "Canal-like" linear features easily distinguishable on Hellas, as well as its brighter appearance and irregular heart-like outline. Much haze around north polar region. **(e)** September 4, 2003 7:30 UT Seeing II-III. Southern polar cap has broken into two parts where fractures were seen in August (mostly ice remaining in the cap, the "Mountains of Mitchel"). Pronounced darkening around cap. Far west third forked extension of Sinus Sabaeus quite readily seen. Northernmost extension of Margeritifer Sinus easily shows small gap within it.

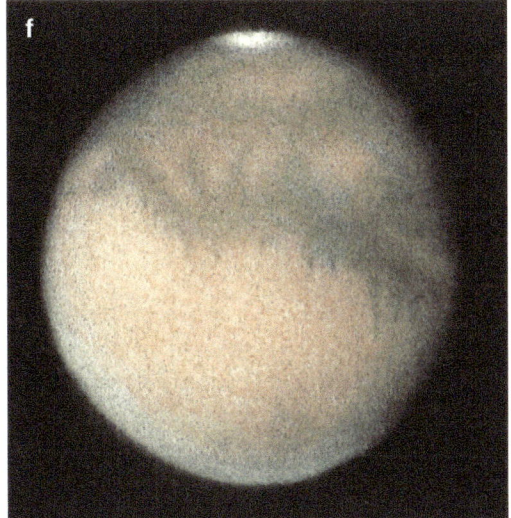

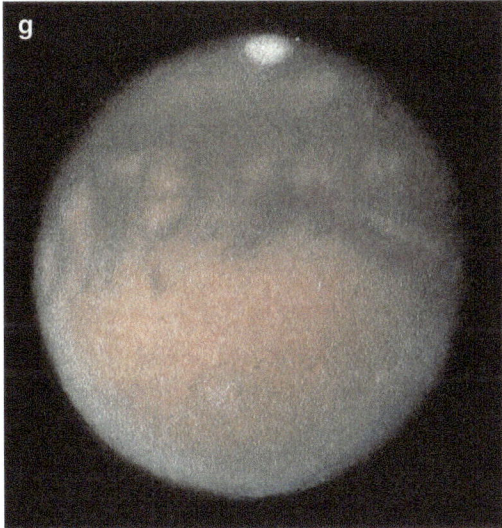

(**f**) September 26, 2003 5:30 UT Seeing III. Incredibly complex detail throughout southern hemisphere. Solis Lacus very prominent this year, with cobweb-like extensions to surrounding encircling ring. Noticeable darkening on western side of south polar cap; could this be a shadow effect caused by it being at a higher altitude? Noticeable phase now on disk. Near circular structural formation of Olympus Mons clearly visible. (**g**) September 21, 2003 6:45 UT Seeing II. Wonderful detail showing throughout equatorial regions, particularly in short and prominent streaks. South polar cap almost invisible; wide dark region encircling mid-south latitudes; some south to north linear detail in these regions. (AC)

conditions, are also clear, recorded as they appeared to the eye at that time. You will also be able to detect limb hazes on most images, both white and pale blue, as well as the Syrtis "blue cloud" effect. The pale form of Olympus Mons even showed well on simple video footage, (f)!

When one has collected many images over a single or several apparitions it is logical and highly rewarding to assemble charts of the surface, gleaning details from everything to achieve the greatest effect and completeness. While compiling a complete chart will take a period of several years (because of the varying axial tilt relative to us), it is possible to produce quite a satisfactory example of most of the surface from one apparition only. In the book *Visual Astronomy in the Suburbs* such a map of Mars was based on a single apparition, that of 1999. This map was drawn from many observations throughout that particular opposition and was made using a Mercater projection, which is one of numerous possible standards to choose in map making. Most of us, familiar with this system as it is used in world atlases, know that it produces increasing extremes of distortion as the latitudes approach the poles in order to lay the projection out in a rectangular, flat fashion for the page. In that map, because of the planetary aspect relative to Earth of the particular Martian approach, the extreme southern region of the planet remained out of view during the opposition and hence was omitted. Such differing aspects are the normal state of affairs with Martian oppositions, and although it might seem to be an inconvenience in some ways, it does allow us alternating and better views of the regions toward the poles (and even slightly beyond) from opposition to opposition. It also allows us to actually see considerably more of the disc from more viewpoints than would otherwise be the case, something we do not experience to any noticeable degree with Jupiter or Saturn. However, it is possible to go one step further, if you have the opportunity:

Quick Project: Mapping Mars

Time Required: 30 min to 2 h, Depending on Your Attention to Detail

Figure 11.2 is a chart of the entire surface, incorporating viewed features from the 1999, 2001, and 2003 apparitions collectively. In this instance, a different projection to Mercater's was selected, one that is also frequently used for full globe charts and maps. Known as a Mollweide projection (specifically, it is of the 'Eckert IV equal area'

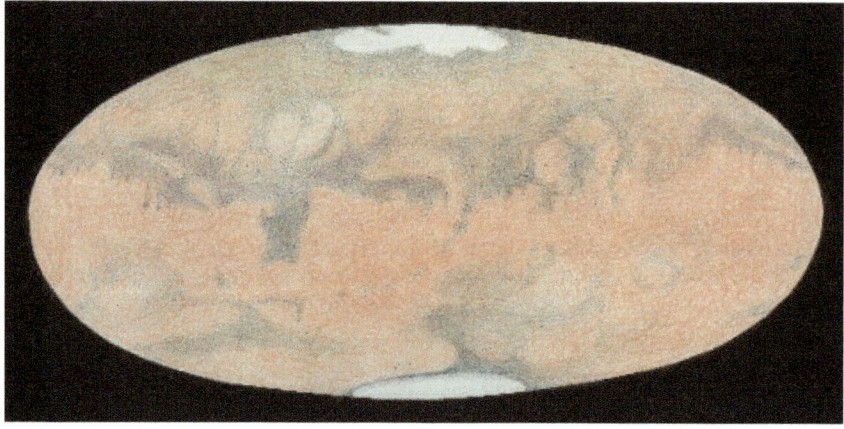

Fig. 11.2. Full Mars globe: compilation from 1999, 2001, and 2003. Drawing. (AC)

type!), it allows a very pleasing full planet image, since less polar distortion of the features is required as we lay it out on the flat page, although the chart itself cannot be rectangular. For the example here, it was possible to complete the chart projection all the way to both poles, and even slightly beyond to peek around the polar caps, revealing more of their full surfaces. Obviously, the oppositions of 1999 and 2003 provided the majority of polar information, since they represented the two greatest extremes.

Draw the features as they appeared during the actual observations on which they are based, to represent their live appearance as much as possible. Because of the choice of projection here, most of the planet's features may be displayed in proportions closely resembling the full disc images. However, it will still be necessary to present features near the poles somewhat expanded laterally in order to incorporate the full global chart.

First, configure and cut out the basic, total, laterally expanded blank from white paper. (The shape here was based on a composite Hubble image of the Red Planet. Since quick, effective results remain the goal, the objective was not to become a renowned cartographer, only to put on the page a good representation of what was experienced at the eyepiece.) Behind this white blank glue a black paper background, and scan the completed template into your computer. There are, of course, far more sophisticated methods available to produce such a basis for your own drawing, but at least you can be assured of a reasonably correct shape, although its symmetry may not be not perfect. In any event, it is an easy matter to print the blank expanded disc on a black background at any scale required, but you might try to approximate your chosen full disc image scale (actually, the dimensions from pole to pole) to allow you the easiest transfer of features to the chart from your full disc drawings.

For such full globe charts, which cover many observations, rapidly dissipating transitory atmospheric phenomena, such as limb hazes or minor dust storms should be eliminated. Changing features should be represented according to your best judgment; in the example here, the southern polar cap is shown as it appeared in early August 2003, and the northern polar cap as it was in May 1999. There was also no attempt made to show any atmospheric absorption at the limbs, and the features were drawn using exactly the same method as in my disc drawings. Naturally, some of the albedo features, common to viewing during these oppositions, did vary slightly over the more than 4-year time frame that forms this composite.

A very good general impression of the entire surface may be easily and realistically obtained by this simple but effective charting method. You will find the results to be deeply satisfying, crystallized into a complete and connected whole. Perhaps they are the most fun you will ever have with Mars. You need not be primarily concerned with producing an exact form of cartography, only a good representation of the appearance of the entire surface, with the generally accurate placement of the features. Greater accuracy, while necessary for true research, is hardly our concern, as we look for ways to maximize our time to attain the greatest results. In that all of ones' own relevant observations for multiple apparitions may be included in one "map" of the entire planet, it would seem to be an ideal way to compress and share the best of our experiences with others. It is especially so when we realize the relative clumsiness of examining the dozens of drawings and video clips that went into this whole. The results will be directly proportional to the time you spend; alternately, you can build the drawing over many short sessions.

We cannot leave this destination without some reference to its two satellites, Phobos and Deimos, although they probably do not rank as candidates for a quick and easy study. Seeing them is a painstaking and fastidious undertaking. Normally invisible, they can, however, be made out as tiny star-like points in certain circumstances, as long as a light blocking bar is placed into the field of view to eliminate the glare from the dazzlingly bright planet they are orbiting. The feat is all the more remarkable if we consider the enormous distance that separates us from them, and their truly diminutive dimensions. If you want to see them for yourself, it is worth the chase, but all of your observing skills will need to be called into play, and you probably will find that time is your enemy. You will also need to do some research so that you know exactly where and when to look. Check out ALPO, the Association of Lunar and Planetary Observers at www.lpl.arizona.edu/alpo.com for information on prime viewing opportunities.

Jupiter

Traveling outward in the solar system to the next of the *big three*, some would argue that the next destination, Jupiter, is the single best celestial subject of *all* for the amateur astronomer. This is a justifiable viewpoint since it is readily accessible for so much of the time and presents a grand and detailed disc in the field of view, with endless variety. For this alone, it seems to deserve its classical nomenclature as the "bringer of jollity," for it certainly brings all of that to the amateur astronomer!

Aside from showing us such a wonderfully large and detailed disc, the great planet provides rapid change, beautiful colors, frequency in the sky, length of apparition, even detail on its satellites; the list goes on and on. Only the Moon and Venus, and occasionally Mars, rival it as the brightest object in the sky (other than the Sun, of course). Best of all, you do not even need a large telescope to see this world in magnificent detail, although you will still need minimums, as set out in Chap. 2. Worthy of much more than being a mere CCD imaging target, seemingly often the case these days, next to the Moon it is the most readily accessible of solar system subjects, and also least jealous of its secrets.

All of this makes Jupiter hard to beat for live viewing; one would be hard pressed to find a subject that can occupy the observer so completely and easily. Indeed, in much the same way that Mars historically has held a unique pull on some notable observers; others, in turn, have found Jupiter so compelling that they were known rarely to observe anything else. B. M. Peek was one such observer. His famous work, *The Planet Jupiter of 1958* (Reprint by Faber) stands out as testament to such devotion, as does Peek's long directorship of the Jupiter Section of the British Astronomical Association. This text still makes wonderful and inspiring reading, even if many of the conclusions have been superseded by more up-to-date information. Certainly the spirit and depth of Peek's enthusiasm makes the pursuits of many more "modern" amateurs seem trivial in comparison. Maybe less sensationally than Mars, Jupiter posed great mysteries and questions for observers not so very long ago. Was there a solid surface lying just beneath the clouds? What was the true nature of the Great Red Spot and other similar features?

Although Jupiter is indeed rich in detail, its many subtleties and refined features appear striking to us not only because the great planet is relatively near, astronomically speaking, but also because it is the largest planet – by far. (It is so large that had it been just a little more massive, it might have become a small star.) Surprisingly, once we reach a certain upward point in telescopic aperture, most of the visible detail has become well resolved in live viewing. It would take a far larger telescope to show dramatically more, but this does not mean that larger amateur telescopes are redundant with this particular planetary subject, because smaller apertures will only show the detail in more muted shades, or even in degrees of gray. Therefore, to reveal infinitely wider spectral ranges on the planet, larger amateur telescopes bring out the reds and browns of the large belts more (along with a variety of dark or light streaks and spots within them), the salmon pink, the darker core of the Great Red Spot, and also shades of blue and other shades in the equatorial zone's so-called festoons (features which resemble large trailing streaks and wisps). Countless similar subtleties of color and detail can be seen across the entire disc.

The Great Red Spot, that huge swirling and apparently indestructible storm, also provides a good example of the advantages of aperture. Through larger amateur telescopes you will *always* be able to see this feature easily, and as a bright salmon pink color. It is also typical to see considerable color shadings and detail within the spot itself. By contrast, in smaller telescopes, it is often described as hard to make out at all when it is at its faintest, and often appears only grayish in tone even at relatively favorable times. As telescopes increase in size, other variations and subtleties throughout all of the individual zones of Jupiter, and the variations in the yellowish overall color of the disc, are more likely to be discernible. It is interesting that the equatorial region varies considerably in yellow intensity, ranging from pale yellow to a more orange tone.

Part of the fun in observing this planet is the constant progression of rapid changes to the surface. Jupiter is so compelling an object that you may be tempted to make whole disc drawings of it at *every* opportunity. However, eventually you may realize that this compulsion to record images, even using the simplest methods, is actually detracting from the time you have as an observer! Over time, as you become more selective in choosing drawing opportunities, you may gradually move away from drawing full discs and move on to detailed close-ups of individual regions, cloud belts, specific features, or whole disc projections. A huge benefit now is that so much satisfaction may be had in so little time.

Quick Project: Drawing Small Regions of Jupiter's Disc and Cylindrical Projections

Time Required: Up to 2–3 min of Observation Every Hour or so

This method of approach seems particularly relevant to our discussion. When you have the restriction of focusing all of your attention on a limited area, you can accomplish a lot rapidly, without having to attend to the entire planet's disc. You will be able to detect and resolve remarkable degrees of detail that might otherwise go unnoticed. Try this for yourself, using the same methods already described for full disc drawing, except that it will be over an extended period. This is an incredible eye training and drafting experience, and you can achieve effective results in just half an hour of total time, freeing you to attend to other matters throughout the observing period. Perfect for the busy astronomer? Definitely so! (Fig. 11.3).

You will notice on the final belt drawing the times that the particular portion was observed; this was a harbinger of things to come. Once able to see the potential that this method of observing offered, as well as the growing ability to make out ever greater amounts of detail, it was not long before sole attention was given to relatively small vertical strips of the entire planet at one time, from pole to pole. This resulted in grand "cylindrical projections" of the whole cloud-topped surface, or at least most of it, during as much of a rotation as possible. Notwithstanding the continual change of the markings of course (!), it is not unlike the effect of a Martian surface map, though with obvious differences in the methods of gathering information.

Again, you can spread out short sessions at the telescope to make up the total. However, you cannot necessarily take for granted a full tour around the planet, even considering its rapid rotation, since it is still possible that it will not be quite rapid enough to complete the rotation! Most typically, it is likely that there will be insufficient hours of Jupiter's optimal placement in the sky for such potential to be realized. By the time larger telescopes have cooled off to the falling temperature of the night air,

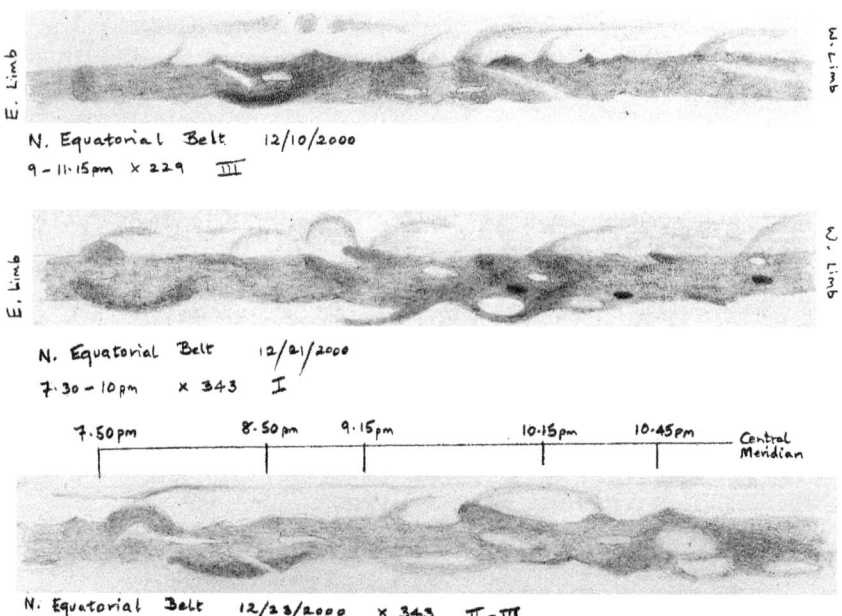

Fig. 11.3. North equatorial belt: one page of detail drawings from the author's sketchbook (2000). Using an 18-in. telescope. (AC)

several hours of potential viewing of Jupiter's path across the sky may have already passed by. Nevertheless, even an incomplete rotation provides an entirely new perspective and insight. Brief periods throughout an evening will be sufficient to produce excellent results, but make sure your observations take place at a reasonably spaced frequency (Fig. 11.4).

It is easy to justify returning to full planet discs, using the newest method of combining video frames and drawing techniques. These images more closely simulate the live appearance than simple drawing, no matter how you are likely to try, and many of the more mundane chores of traditional drawing are eliminated.

A great advantage to this approach is the leisurely way you may allow yourself to go about gathering your basic images. Because of Jupiter's rapid rotation, a drawing from scratch requires considerable skill, as well as organization, in order to accurately portray all that is visible before significant rotation has taken place. If you have spent any time at all with Jupiter, you will realize just how fast its rotation is. And it is easier said than done to place and scale the belts correctly. Video frames, however, along with rapidly made supplementary sketches and remarks, effectively eliminate the problem, allowing for a very accurate and timely representation to be made at a later time. Perfect for our circumstances! The basic printed video frames also show relative shadings of the disc and effects of limb absorption (though somewhat more so than live through the eyepiece), as well as precise belt and feature placement.

Jupiter is always a source of great interest, with surprises in its appearance occurring regularly. Even when the planet seems relatively "quiet," we may rest assured that substantial changes will soon take place, although there is always something of interest to be seen. The opposition in 2003 was not particularly spectacular for revealing complex detail in the usually busy equatorial zone or even the strongly visible minor cloud belts. However, the equatorial belts were bold and prominent, and the Great Red Spot stood out well. A series of curious white spots formed an "eyebrow" south of the spot, and it was interesting to watch it evolve, along with the distortion of nearby belts, which normally ride parallel to the equator.

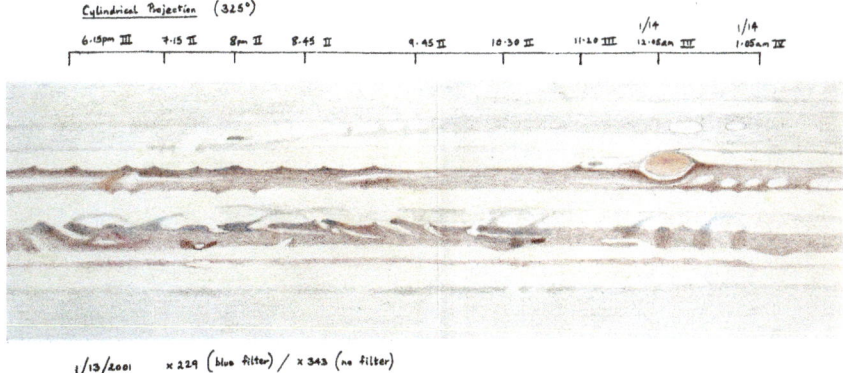

Fig. 11.4. Jupiter: cylindrical projection from the author's sketchbook (2001). Using an 18-in. telescope. (AC)

Later (Figs. 11.5 and 11.6) is a representative sampling of images from the 2003 opposition, all single frame video images combined with drawing. In groups of three, they are full excerpted pages from the author's sketchbook. Fig. 11.5 presents interesting aspects of the Great Red Spot and its surroundings. Again, that elusive quality of luminescence seems to shine through on these images.

In Fig. 11.6, the first image shows the wake of the Great Red Spot, splitting the southern equatorial belt after it has crossed the face of the planet out of view. It is fascinating that this ancient storm is able to have such an enormous influence along

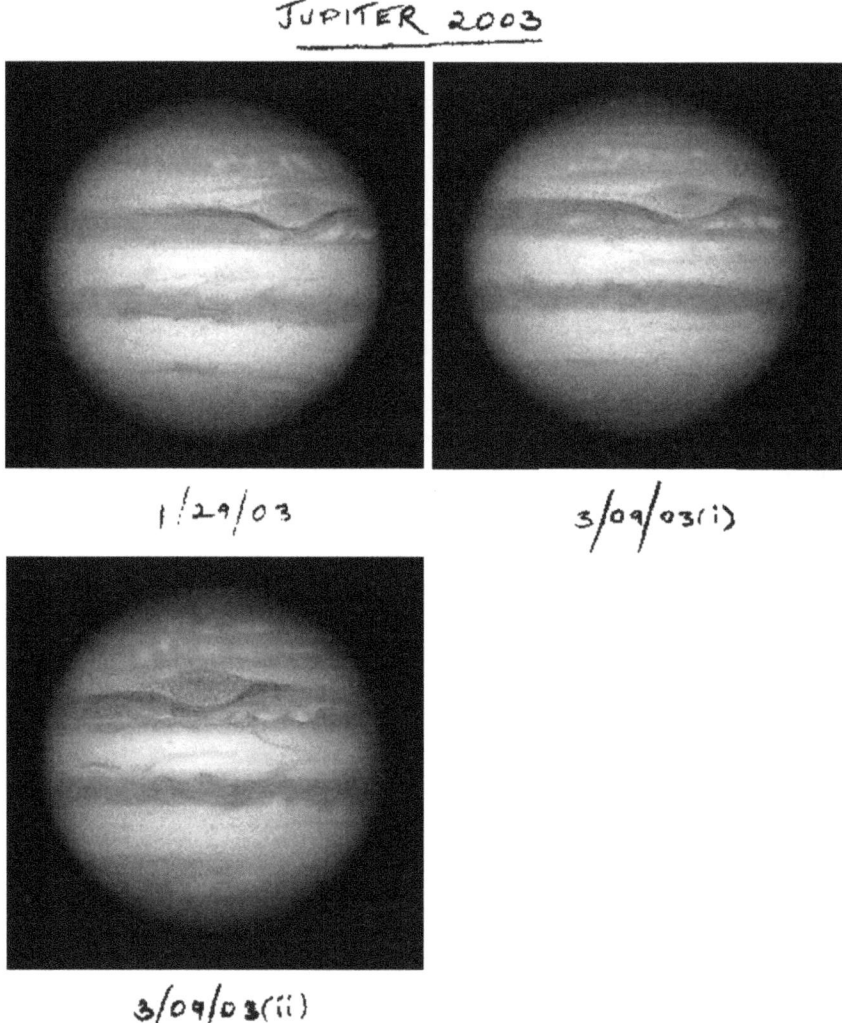

Fig. 11.5. Jupiter 2003: the Great Red Spot. (AC)

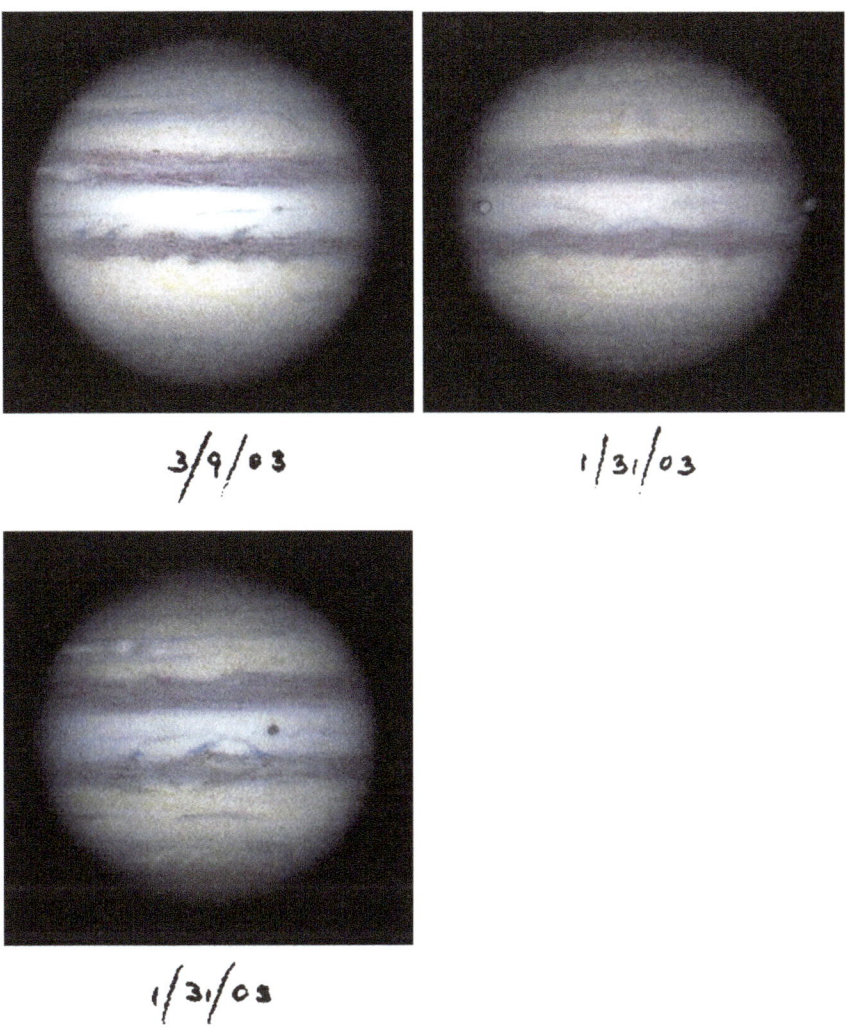

3/9/03

1/31/03

1/31/03

Fig. 11.6. Jupiter 2003: in the wake of the Great Red Spot. (AC)

the belt so many tens of thousands of miles past it. Beyond this, as later images show, the belt finally returns to normal.

Because of Jupiter's proximity to the Sun (compared with the rest of the outer planets), substantial thermally induced interaction of the gases of its atmosphere is constantly taking place, which produces the colorful and changeable detail we see. Jupiter's great variety of appearance may often be detected over very short time periods, and its period of rotation is so fast that it is detectable in minutes. Just one more advantage for the busy observer!

Although it is possible to actually resolve surface detail on the Galilean moons, doing so has challenged many observers over the years. However, if the detail is there to see – you will need exceptionally steady air and high magnifications – it will be readily apparent and immediately obvious with sufficient apertures, which may make the quest all the more appealing, considering the nature of time-compressed astronomy we are pursuing. The lesser known of the two nineteenth/twentieth century Pickering brothers (William, a close associate of Percival Lowell, and not to be confused with his highly distinguished brother, Edward) passionately believed that these moons were oblate in shape, stubbornly clinging to his views despite all evidence to the contrary. The illusion that fooled him, was, in fact, created by the light and dark markings on the satellites themselves, seen against the bright planet behind. His telescopes simply were not of sufficient aperture and resolving power to contradict his beliefs. We must bear in mind once again that, according to Dawes Limits, resolution of detail on these tiny worlds is theoretically impossible with the apertures Pickering had at his disposal, and those usually available to amateurs, even today. However, when seen against the disc of the planet itself, the bright contrast of tiny moon against stark black sky is eliminated, and the slightly contrasting backdrop of the huge planet's less bright limb provides us an opportunity to see something on these moons.

Although enhanced electronic (CCD or similar) images have revealed striking amounts of detail on the Galilean satellites with only moderate apertures and careful processing, you should not count on similar success live at the eyepiece. To make out some surface details, you will need to see them through telescopes of around 12 in. (30 cm) and above. And bear in mind that as each satellite continues its journey across Jupiter's disc toward its center, it will encounter a brightening of the planetary background, so plan your observations carefully to take place early on in the transit. Regardless, it is always easy for any of these moons to become completely lost against this background. Without undertaking extensive observations of the moons you may be startled at times to see dark shadings, and at times even specific traces of color on them. This is especially true of Ganymede, the largest of the Galilean satellites. More than ever, for your best chance at seeing detail on the satellites, you will need to wait for your best viewing conditions, when the great planet flops and wobbles only slightly in the field of view. These conditions will not occur too often, although when they do, it is truly wonderful to witness such clarity and with such ease. These are the occasions that will always stand out in your memory, when typically evasive planetary detail also presents itself almost as if it were imaged from a spacecraft; however, it is happening right in front of your eyes!

Saturn

There is probably no sight in the whole universe better known, or perhaps more compellingly beautiful, than mighty ringed Saturn. Often a beginner's first sight through a telescope, and one of the few celestial objects guaranteed not to disappoint the uninitiated, there is nothing quite so mesmerizing as this spectacle. It is certainly the planet most traditionally employed to conjure up visions of outer space, even if frequently it is portrayed inaccurately. Never mind; even outsiders know which planet is being conveyed.

Unfortunately, with repeated viewings, even with pretty big apertures, we realize that the kind of complexity and variation we find on Mars and Jupiter is somewhat elusive on Saturn. Named as the "bringer of old age" in classical mythology, perhaps we feel as if we will die of old age before we see much change here! Nevertheless, telescopes of the highest quality and decent size will help in revealing whatever changing detail there is to be seen from our distant vantage point. The ringed planet is large enough to be quite impressive in the field of view even with relatively low powers. However, the width of its diameter *together with* the full width of the ring system is barely more than that of Jupiter's entire disc at the same magnification. Therefore, we are already at a distinct visual disadvantage, though this is really only true when compared with Jupiter!

Certain features are immediately obvious with the slightest optical aid. Cassini's Division, the dark gap in the ring system (a small feature in the great scheme of things), is a remarkably easy mark in small apertures with only modest magnifications, because of its high contrast relative to the rest of the ring system. However, larger and better telescopes will always help in bringing out many other subtleties on the disc and rings as well, although there is no way to actually see more than a tiny fraction of the near countless subrings and divisions of the beautiful ring system, all held in perfect balance in orbit around the planet itself. Spacecrafts have recorded hundreds of similar divisions and segments. Interestingly, Saturn can often stand seemingly reckless amounts of magnification (good optics and conditions being a given, of course), even more so than the other members of the *big three* fraternity. These kinds of powers would be sufficient to ruin the definition and contrast on practically anything else. Try it out; you may be pleasantly surprised, but remember such results will only be possible when atmospheric stability permits; nothing is worse than pushing the magnification beyond the conditions or the capabilities of the telescope you are using.[1]

With decent apertures, even on nights of only moderately good air, you will be able to make out many refinements of detail. At best, these features again may remind you of Hubble Space Telescope pictures, with the exception being that there does not seem to be any way of imaging or even describing all of the extreme subtleties, even by drawing. The question of determining exactly what we are seeing remains our greatest challenge; we know we can see it, if only we could be sure what it is! Saturn, in some ways even more than Mars or Jupiter, seems notoriously fickle with its details. Naturally, ever larger telescopes make this task somewhat easier, and also will bring out more of the colors present on the disc as well. Along with the dominant and almost indescribable golden hue of the disc and rings, the deep red and brown hues of the equatorial belts, the darkening and zones (even shades of green) in the polar regions, seasonal white spots, and pale equatorial zones are all things to look for. And one never knows when a great spot (typically pale colored) will appear. Although relatively rare, they are not unknown, and examples of spots of at least moderate size occur fairly often.

---

[1]Many years ago the magnification 710× was put to dramatic and worthwhile use on Saturn with my old home built 12½-in. (31 cm) reflector (Fig. 2.1, Chap. 2). Despite the seemingly ridiculous amount of magnification, there was no suggestion of image breakup; of course, the viewing conditions were nothing less than remarkable. The view of Saturn was even better, and spectacularly big in the field of view.

Fine detail on the rings presents numerous other possibilities to resolve at the eyepiece, but you should also recognize the difficulties and limitations that they pose. We may be able to see the bluish-gray hue of the Crepe Ring, two or three differing color intensity zones in the yellow "B" ring, and a couple of other zones in the darker "C" ring. Maybe you can even glimpse some of the now famous transient "spokes" in the "B" ring on occasion, although they do not usually come across on extracted video frames taken during observations. These unusual and seemingly unlikely features have been commented upon for generations, although it took the planetary probes of the space age to decisively confirm their existence. Exactly what causes them or constitutes their makeup has not yet been fully determined, although there is apparently a connection to the planet's magnetic field, or other electrostatic forces. Do not expect them to be obvious. However, with larger apertures there are certain occasions (you will probably need the rings to be fully open to have any chance at all of seeing them) when you will swear there is something about the appearance of the rings that does not seem completely consistent. Appearing something akin to conical dark shadowed areas, if you become aware of such features you are probably seeing traces of these *spokes*. Once you know of their existence, you may be able to positively confirm, at least in your mind, that you have actually seen them. There were more traces of them in 2004 than usual, so they might now figure high on your own list of things to look for.

Aside from the better-known Cassini Division, within the "C" ring is the infamous and much narrower Enke Division. Parts of the "C" ring are of a lighter shade, which can fool you into believing that you are seeing inside the division. Resolving this remarkably fine feature is not quite along the natural order of things that many would have you believe. In fact, it is shocking just how routinely the sight of the division is described in "official" amateur circles, especially considering that its dimensions are actually well beyond resolution in all but the largest amateur scopes, and only when they are used in the very best observing conditions. Even the great E. E. Barnard (generally considered to be the most gifted visual astronomer of all time) failed to detect it with the mighty Lick 36-in. refractor. And Barnard had seen *spokes* on the ring system! The Enke Division's discovery had to wait for many more years. Most observers cannot honestly report having seen it before having the advantage of knowing it was there, and certainly not without the benefit of larger apertures. However, it is *still* beyond the theoretical resolution of most of them! The fact that it can sometimes be seen – or more likely, glimpsed – in lesser apertures than the optical limits would seem to allow is because of its high contrast relative to the bright rings. Dawes can be superseded at times!

A jaw dropping statement recently in a prominent and respected magazine that "about 8 inches is needed in order to show Enke's Division" was among the biggest overstatements ever seen in such a respected source. How many readers are left thinking they have made a poor investment when their scope fails to show any trace of it? How many readers must believe they live in an area of poorer seeing conditions than the norm, or have bad eyes? Worse, how many observers convince themselves they can actually see it? Most likely, telescope users (of moderate to smaller apertures) who report seeing the division are really witnessing the makeup of the darker zone within "C" ring, which is on the inner side of the Enke Division. This region is very hard to describe, even when seen to the best advantage, and seems to consist of tantalizing shreds of detail dancing at the very edge of resolution. These finely spun details are often mistaken for Enke's Division itself, if indeed anything is seen other than just the entire "C" ring! Just look at any Hubble Space Telescope image of Saturn, and you will begin to realize just how fine a feature the division is. It is the finest thread imaginable. Couple that with the small, quivering total image size you will usually have in the eyepiece by comparison, and it is pretty clear that one should not take seeing this feature for granted. Moreover, from an optical standpoint, easy and routine sightings with most amateur equipment are simply not possible. Only the larger sizes have any chance at all. With so many regular reports of Enke sightings, think once again of the purely fictional "canals" of Mars, which were routinely "seen" before their existence was disproved. And some of these same people will tell you Lowell was a crackpot.

The best chance of seeing anything approaching the kind of detail you may anticipate on the rings is when they are presented at their most "open" inclinations. These times occur every 15 years or so, as we progressively see either more of the "top" surface of the ring system, or the other extreme "underneath," in alternating cycles. In the interim, as the rings swing ever closer to an edge-on position, the opportunities to see detail on them naturally diminish, although they now obscure less of the disc itself. These are also now the best chances to compare details on the planet's disc itself to that of its big brother, Jupiter. The equatorial regions are always worthwhile places to watch; occasionally, quite radical lightning storms occur in them. Again, with Saturn, filters do not seem to be especially helpful, although a blue filter certainly makes it look very beautiful, and may bring out a little more of the reddish color of the belts. Note also how the shadow cast by the planet on the rings sometimes appears irregular and angled because of the stark optical dominance of Cassini's Division.

Saturn's apparently static countenance could lead one to deduce that, apart from the attitude of the rings, little is to be gleaned on the surface between adjacent oppositions. On first glance, this would certainly seem to be the case, but a little closer inspection reveals that things are indeed undergoing more change on this planet than is immediately obvious. Detail is always somewhat faint, but by looking carefully in the eyepiece over time you will likely notice differences in the cloud belts, even irregularities in them, or a white spot or two. Occasionally, one of these spots flares into sizable prominence, dominating the landscape for weeks or even months. Figure 11.7, taken about a year later than the example shown in Chap. 10, p.124 shows that not only have the rings "peaked" in their open aspect (note the near absence of projected shadow on the rings) but also interesting changes to the entire globe have occurred, subtle though they may be. In particular, the north equatorial belt was brighter and sported some striking kinks and irregularities in 2004, even spots (?), compared with the smoother and more regular appearance of the previous apparition. North of this belt, the planet had begun to appear paler with a less prominent belt, and in the opposite direction, south, other fainter traces of belts were to be seen.

Other subtle differences in coloration from image to image are possibly due to atmospheric effects, simultaneous contrast, and variations of your own subjective judgment. However, the unique golden coloration of the entire planet and rings seems more realistic in these representations than in those made by most other imaging methods. Even the best of CCD imaging usually seems to make the planet appear too white and varied in color. Hold the page at arms length, and this effect becomes even more like the appearance of the planet in the field of view! Compare these images against your own viewing, and determine for yourself just how successfully

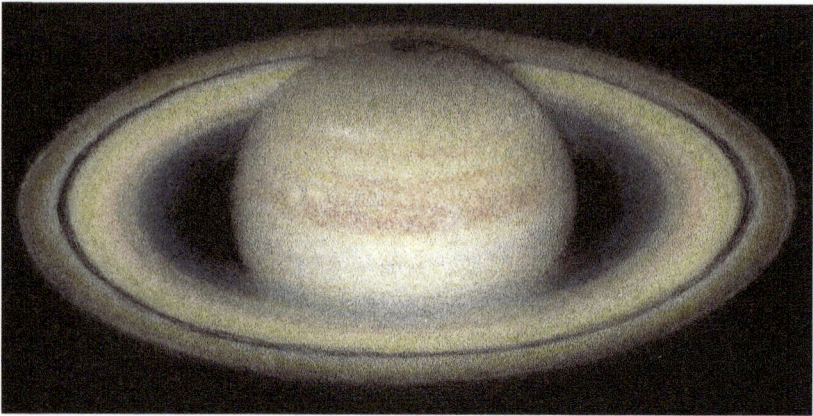

Fig. 11.7. Saturn January 17, 2004. Combined video frame and drawing; an 18-in. reflector.

(or not!) the essence of this planet was captured. In fairness, numerous very fine CCD (and other electronically made) images of Saturn do exist, even if they do lack the unique golden brilliance.

Unfortunately, Saturn's relatively undemonstrative state does not provide us with nearly the range of projects to undertake as do Mars and Jupiter, such as multiple close-dated drawings. Most individual features, when seen, will not usually present sufficient detail to merit separate images, something equally true for making cylindrical projections or belt drawings. Aside from Saturn's relatively quiet state, its reduced scale compared with Jupiter is also partly responsible for the challenges we face. More the pity, but at least the prospect of a sudden and dramatic arrival of a spot or spots, or the prospect of detecting *spokes* in the rings may give incentive to return again and again, if the amazing spectacle of the ringed planet in itself does not! But there is one other unique pleasure Saturn holds for us: the endless display of changes in the aspect of the rings themselves is always exciting to see for oneself.

Unfortunately, Saturn's satellites do not provide the opportunities we have been treated to with Jupiter's famous four. They are simply too distant and too small for such viewing, and any chance of seeing even the form of a disc against the planet is not in the cards, although many observers enjoy watching the ever-changing positions of the brightest members of the system. However, it is still interesting to watch their positions in orbit around the great planet change constantly, just as with Jupiter's moons. Spacecraft have shown us many wonderful views of these remote little worlds, with even the tempting thought that some of their environments could harbor some forms of life. We will not know the truth about this possibility, of course, for many years to come.

For us, one of Saturn's best attributes is that most of what it has to show on any particular night may be seen almost immediately and in short observing sessions. This would seem to be just what the doctor ordered!

CHAPTER TWELVE

The Far In and Far Out

With the three grandest planets in the solar system always likely to remain a major focal point of the amateur's visible universe, there are nevertheless some other worthy sights in the local neighborhood. It has to be said, however, at least in the visual sense, that most observers will always regard these other destinations as the solar system's "poor relations." As potential spectacles for live viewing, except for the occasional appearance of a great comet, we must reluctantly concede that indeed they *are* the poor relations! Nevertheless, there is still much to enjoy and explore as long as we approach them with just a little different attitude. However, especially with time not on our side, it would be useless to pretend that, at the telescope, these subjects are comparable in any way to those of the last chapter, or that equivalent success may be achieved with them. Rather, we need to approach these elusive objects with a special appreciation of what we are actually seeing.

We need to take advantage of the insights gained from closeup rendezvous of space missions, or images obtained with giant earthbound telescopes fitted with adaptive optics and other enhancing equipment. Thus, seeing them again through the telescope will be a different experience. What we are witnessing for ourselves will never be more meaningful, and less likely to seem uninteresting, when compared with the more accessible and more visually dramatic places of the solar system. Visit the NASA web sites listed in Chap. 16; in the regions of the solar system described within this chapter, these sites play a larger role than ever.

The remaining planets, certainly less accessible worlds, comprise those inside Earth's orbit and those outside that of Saturn's. They have always posed problems for viewing from Earth. Even when we do find them in our telescopes' fields of view, they do not present particularly impressive detail, at least anything that is immediately obvious. However, while not exactly *spectacular* objects in the normally accepted sense of the word, there are still good reasons to visit them and give them at

A. Cooke, *Make Time for the Stars: Fitting Astronomy into Your Busy Life,*
DOI: 10.1007/978-0-387-89341-9_12, © Springer Science+Business Media, LLC 2009

least some part of our attention at the telescope. With a degree of patience and persistence, together with all the other usual prerequisites, the enthusiast even may find them to be more worthy than expected. However, you should not take this to mean that these planets are easy choices for study, and especially for casual observing.

The "Far-In" Planets: Mercury and Venus

The problem of diminutive disc size is always an issue here, and although Venus is often well placed in proximity to Earth, when it does present considerable size, it is seen only as a thin crescent! The placement of Mercury, being more distant, has similar problems as well as being much smaller than Venus. Being so close to the Sun limits its time above the horizon, and seldom is it observable for long after sunset or before sunrise; it is also never high in the sky. In a strange quirk of fate, both Mercury and Venus make wonderful naked eye objects and profoundly disappointing destinations in the novice's telescope.

When their orbits bring them to the same side of the Sun as Earth, as long as they are not directly between it and us, they appear as crescents, varying substantially in size and shape, depending on their exact placement. However, to have any chance of discerning any detail, we will need to have something closer to a full disc to observe. Sadly, as planets positioned inside Earth's orbit, they only present such discs when on the opposite side of the Sun from us, since this is the only placement that allows full illumination from our viewpoint. Of course, you may have already deduced that when they are farthest from us, they appear much smaller in the field of view.

All in all, to most casual observers the inner planets are disappointingly difficult and frustrating objects. This seems all the more regrettable since they are so close, astronomically speaking. To have any chance of resolving detail of any kind, high magnifications are therefore required, and your telescope needs to be well adjusted relative to the temperature of the surrounding air. This is often difficult to attain because of these planets' appearance in the skies so close to sunset, when our telescopes are struggling to attain thermal equilibrium. With moderate to large amateur telescopes, cooling to the surrounding temperature is always an issue. Using smaller telescopes to combat transient thermal issues can help, but then we lose the advantages of larger ones. Climates with minimal temperature swings therefore will probably be most generally favorable. For the rest of us, unless the temperature drop-off at any time is likely to be slight, you may be better off waiting for appearances in the early hours of the morning instead, when your telescope has had many hours to adjust. All of this may be enough to deter you from trying, since there are more readily accessible destinations to spend a little time. You must decide for yourself the merits.

The various problems with viewing the inner members of the Sun's family have been extensively documented over the years. Venus is always shrouded in mystery under a thick cloud blanket. Seeing certain patches of markings is possible, but never an easy task. These patches do not, however, correspond to actual features on the surface, and although these cloud formations can indeed be seen, you should be aware of the challenge they pose. Mercury is no more generous, despite having no such cloud cover at all. It is a scorched dead world, cratered much like the Moon, but far too small and distant to allow us to make out more than the slightest patchy surface markings and never anything of its true nature.

Additionally, viewing both Mercury and Venus are further compromised not only by their early disappearance from the evening and morning skies but also the infrequency of their favorable, or even viewable placements, most especially Mercury. Needless to say, even when air temperatures and stability are relatively suited to successful viewing, all of the other drawbacks in viewing both of these planets remain. Because these two inner planets will always follow or precede the Sun quite closely, for Mercury it always precludes a good overhead placement in the sky, although sometimes Venus fares better. This almost guarantees less than ideal atmospheric stability because of the need to peer through the densest layers of air. And remember, when they appear brightest and at the highest possible placements in the sky they will appear only as the narrowest of crescents!

And yes, there is *yet* another problem. Both of these objects are also so glaringly bright in the eyepiece that the resolution of any potential features is always difficult, especially in live viewing. Best results will probably be obtained by viewing both planets in a less than fully dark sky (something that comes with their territory), and the use of polarizing filters is recommended – violet, or even Moon filters to help cut back the glare. Regardless, it is one of astronomy's greatest disappointments that such brilliantly visible objects reveal so little through the telescope, or at least, easily so. But let us look a little further.

Mercury is so insignificant in size (3,100 miles/4,988 km in diameter) that its great distance from us, relative rarity in the skies, plus its low placement keeps it a highly elusive object to study. Once you do manage to train your telescope on it, do not expect to see more than the tiniest image in the eyepiece field of view, or more likely a featureless crescent. During initial observations it is unlikely that you will be able to detect any surface features at all! However, with a little patience you may be able to make out some vague dusky markings once in a while, including larger variations in brightness across its disc. Mercury also shows a distinctive off-white or pinkish color, due more to its close proximity to the setting Sun's reddish glow rather than its intrinsic color. Most observers find it extremely rewarding just to see it once in a while, and discerning such markings as there are to see. Amazingly, many casual observers have yet to positively identify it, let alone study it through a telescope. This is especially true from city locations, where the horizon is all too often obscured by buildings, along with Mercury's fleeting presence in the twilight skies. However, it is surely worth spending a half-hour observing session once in a while if your circumstances allow it.

Our knowledge of the surface of Mercury was forever changed in 1974, when *Mariner 10* was first to swoop by its surface, and now again, and more dramatically by the *Messenger* spacecraft mission of 2004 through the present (2008). Unsurprisingly, the "winged planet" was seen to be an arid and parched little world, appearing not unlike the surface of our own Moon, complete with craters, light and dark shadings, and even some smooth plains! Figure 12.1 helps provide a little insight before you attempt to observe it. However, despite a superficial resemblance to our own Moon, it is in no way possible to see these qualities for yourself through the telescope.

Fig. 12.1. Mercury, March 29, 1974, imaged by *Mariner 10* (photocourtesy of NASA/ JPL-Caltech).

If you still have a little "solar phobia" with Mercury, it is perfectly understand-able. Because its position is so close to the Sun, often shining through the bright after-sunset glow, you may find it hard to overcome an innate resistance to pointing your telescope anywhere near the Sun, even after it has set! It is possible that only when the "winged messenger" is at higher positions in the sky will you find yourself venturing toward it. Crazy, perhaps over cautious maybe, but at least you will still have your eyesight!

Venus is much larger than its nearby brother, being comparable in size to Earth. But before you get your hopes up too high of seeing dramatic Venusian landscapes, remember that it, like Mercury, has many of the same challenges as well as different ones. Many years ago, there was a raging debate about exactly what, if anything, was visible beneath its thick (mostly carbon dioxide) cloud cover. Even its rotation remained shrouded in as much mystery as the nature of the surface. The French amateur astronomer and dedicated Venus observer Charles Boyer (not the actor!) actually solved some of its major riddles (although it was many years later that his findings were finally acknowl-edged and confirmed) and was, in fact, able to study some actual "features" of the cloud cover.

Fortunately, Boyer was taken seriously by some and was able to have access to France's Pic du Midi Observatory. In this way he was able to capitalize on superior equipment and the near perfect planetary viewing conditions frequently experienced at this high-flying site. Nevertheless, do not for a second think that he had an easy task in what he had set out to do! Boyer was the first person to correctly establish the thick cloud cover's unlikely 4-day retrograde motion. This period seemed much too fast, relative to what was known of the planet's similarly retrograde rotation (at 243 days), and was not accepted in the astronomical community for many years. His achievements are still remarkable and served as a benchmark as to what can be attained by developing the highest levels of viewing skill and patience. For most of us just trying to keep up with life, such efforts would be out of the question! Only when spacecraft confirmed Boyer's findings was the true magnitude of his work grudgingly accorded. *Sky & Telescope* magazine published (June 1999) a detailed account of Boyer's work; any self-respecting Venus observer should try to read it.

The uninitiated might be forgiven for believing that these features seen by spacecraft are actually present on the surface itself, which it is peering through the clouds. But in fact, they are all common to the top of the cloud layer exclusively. The now famous "Y" shape and other identifiable dark features clearly seen when photographed in ultraviolet light are actually variations in the cloud layer and belong to the equatorial regions. Even "polar caps" of sorts can be discerned on these images, but they are again, in fact, caused only by the actions of the atmospheric currents and temperatures, swirling differently around the planet's waist than at its poles. Visually, it is sometimes possible to make out something of these "caps" with relatively modest equipment, but only when the planet is large enough in the field of view. This potential is eliminated, however, when it appears as a slim crescent (and hence is closest to us), because we need a reasonable portion of Venus' disc to be visible to see the dark markings or *caps* imposed upon it.

Quick Project: Viewing Cloud Detail on Venus

Time Required: 10 min

Most of the time, little or nothing can be made out on the dazzling white disc itself, and for really detailed views of the cloud layer together with any suggestions of features within them we have to depend on images made in ultraviolet light. However, you might try viewing the disc through a Wratten 47 dark violet filter to see something of the variations in the clouds. It is indeed possible to make out darker and brighter regions in this way, typically the V-shaped formations revealed in image Fig. 12.2, if you look carefully enough. Some experienced observers have also commented on the granular appearance of the cloud covering when viewed in outstanding conditions.

Although the image (Fig. 12.2) actually provides a good pointer to the nature of what you may *just* be able to see for yourself at the eyepiece, you will need extremely favorable conditions and, of course, a violet filter. But please realize that it is more "glimpse" than "see." Do not expect such dramatic clarity and detail, although you can hope to see in this image the essence of what you are looking for: the "polar cap" features as well as dark shadings around the equator. You may also catch a glimpse of the so-called *Ashen Light*, the unlit portion of the disc itself. Many observers are familiar with this phenomenon, which looks something akin to "Earthshine" on the

Fig. 12.2. Venus, February 5, 1979. Imaged in ultraviolet light by the *Pioneer* Venus orbiter (photo courtesy of NASA/JPL-Caltech).

unlit part of the lunar disc. Its cause may be anything from Sunlight illuminating the clouds to electrical activity within them. Although Venus is at best difficult to image successfully, you may nevertheless sometimes be able to record something of the geometric cloud shapes so often seen in images such as these, again with a violet filter. However, going through this process and all that is entailed may take more time than you have.

The actual surface itself always remains invisible to us from Earth. An imagined inhabitant (no chance, with such searing temperatures!) might never gain knowledge, let alone awareness, of the universe, the skies above remaining in perpetual gloom. Visually, the little that we do know of the scorched and shattered surface is as a direct result of visiting spacecraft, and the only one that successfully reached the surface (Russia's *Venera 7* in 1970) transmitted information and imagery for only a short period of time before yielding to the intense, blast furnace-like heat experienced at ground level. All in all, the classical planet of "love" turned out to be none too lovely; more like hell in fact! All other knowledge of its topography has been gleaned from orbital radar, which has been quite successful with the maps produced. As far as features are concerned, Venus is indeed a very bland and hostile world, indicating none of the actions of water to carve out its face and generally showing remarkably little variation of altitudes. It is unlikely that you will lay claim to spectacular visual results in viewing or imaging the cloud tops of Venus, but you may have some success in seeing at least some of what is outlined here. Worth your efforts? Definitely. But as always, when time is a factor, you must decide when to try. In any event, just knowing what is potentially visible will add much to your pleasure and productive use of your time.

The "Far-Out" Planets: Uranus and Neptune

It is not any easier as we move outward from Saturn's orbit in the solar system! By the time we reach these outer planets, there are ever-greater challenges as the shrinking grip of the Sun corresponds to increasingly far-flung orbits. Although Uranus and Neptune are sizable objects, being so much further away they are quite faint and diminutive in the night sky. And despite being casually similar in their gaseous nature to Jupiter and Saturn, they are of a substantially smaller scale. Compounding the situation even further is the lack of observable detail. This would be the case even if we were situated close to them; the decreased temperatures this far from the Sun produce noticeably less mixing and swirling of the outer gaseous mantles of these massive globes. This, in turn, keeps them quieter and far less colorful worlds.

Despite the diminishing returns from our efforts at the eyepiece as we travel increasingly to the inward and outward reaches of the solar system, there is still reason to spend a little time seeking them out. Although Uranus and Neptune are both gas giants of the Jupiter model, they are completely different in their telescopic appearance. Without the violent activity and colors (as with Jupiter), they make for far less profitable viewing. However, both of these planets have wide optimum viewing windows, and the prospect of making out more than just the discs themselves, whether moons or just a hint of planetary detail itself, remains tantalizing. Just to see these remote worlds in the eyepiece is always a thrill. Because of the great width of these planets' orbits, their oppositions result in their being more favorably placed for viewing by only small amounts, and so for much of our year they present virtually the same disc size to us. The opportunity and possibility of seeing something out of the ordinary is thus always present.

Meanwhile, imaging them by any means remains highly challenging. This is true with all of these "second tier" planets at the best of times; simple CCD video frames may well be insufficient ever to show anything other than the outlines of these planets' discs against the black vault of space. The challenge is not greatly alleviated by more sophisticated means.

Uranus, the closer and larger of the two, with a magnitude approaching 6, while showing a blue-green disc, offers limited opportunity to see anything at all resembling detail. Most observers report that they have never seen any detail on the planet, other than its ghostly and predominantly greenish color evenly spread across its

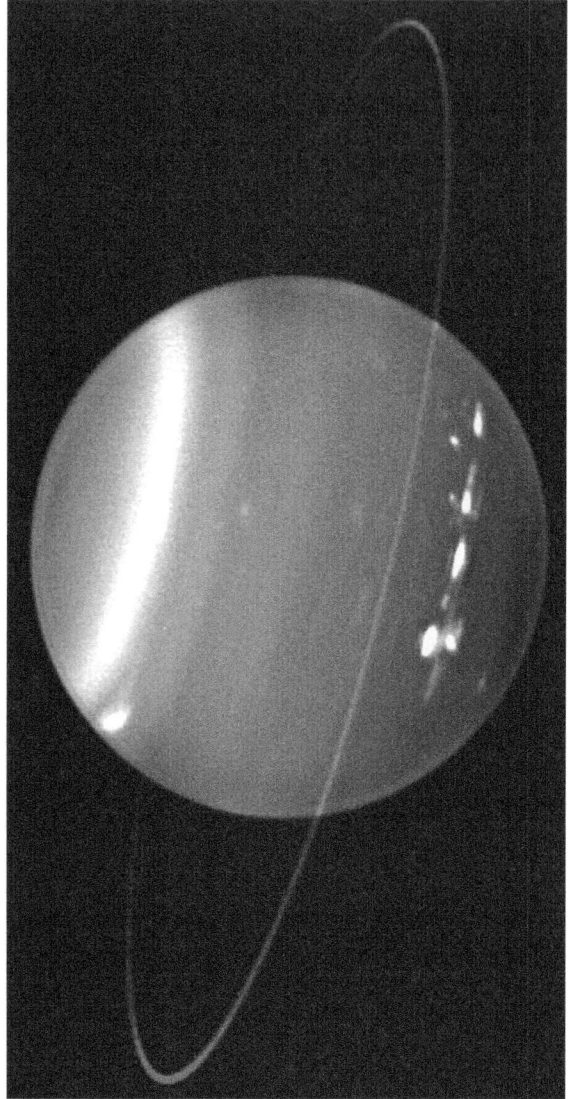

Fig. 12.3. Uranus, July 9, 2004 (Axial tilt results in the N. pole appearing at "4 p.m."). Keck telescope image (photo courtesy of NASA/JPL-Caltech).

entire face. This was even true for the Voyager spacecraft that flew by the planet in 1986! Larger amateur scopes may reveal some of its moons, however. Again, technology has come to our aid in revealing what is impossible at the eyepiece (Fig. 12.3). With newly developed adaptive optics, the Keck Telescope in Hawaii was able to make out some belt and storm activity in 2004; finally, we have something more than a blank stare looking back at us. More compelling still was the well-defined ring system around Uranus's waist. However, do not think for a moment that this ring system compares in any way to that of Saturn! In fact, the rings themselves are quite dark in hue, more like the darkest of gray shades, and would give us nothing of the spectacle that Saturn has provided, even up close; they are certainly not visible to amateurs.

Even with the very best of imaging techniques, Uranus is likely to remain a difficult object, although it is always pleasing to see the planet's substantial disc for oneself. In a short session once in a while, Uranus will usually justify the effort. And thanks to spacecraft and modern telescope technology, the "magician" of classical lore has at last been robbed of its sleight of hand!

Neptune, just like Uranus, is no more likely to provide obvious visual rewards. True to its classical designation, Neptune remained the "mystic" until late in the twentieth century. Being smaller and still more distant than Uranus, it has only a magnitude of 7.8. Its tiny visible disc (at only 2 s of an arc) will present a challenge at the eyepiece for most observers in any conditions. Now, such diminutive image size is usually considered virtually of no value when observing infinitely more detailed and contrasted objects, such as Mars, so you can see what you are up against. Without a sizeable telescope, it is unlikely that you will make out much, but nevertheless what is apparently the true outermost planet would seem to warrant some effort, if just to say you have seen it! The planet is sufficiently faint that you might try observing it with an image intensifier eyepiece, if you have one; its brightness, or lack thereof, certainly corresponds to many deep space objects. At the very least, this

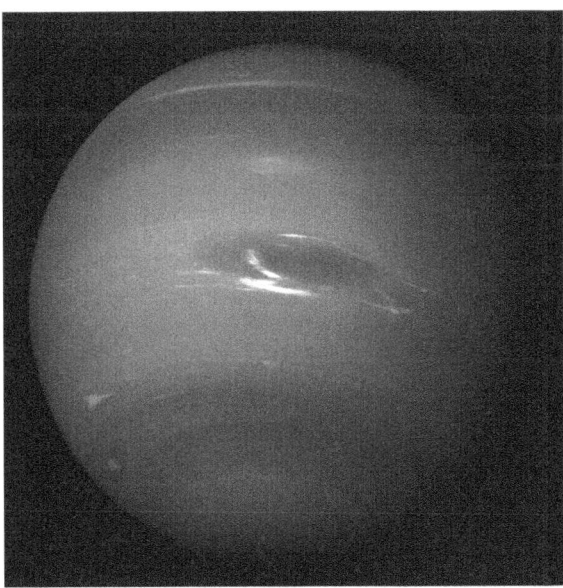

Fig. 12.4. Neptune, August 20, 1989. *Voyager 2.* (photo courtesy of NASA/JPL-Caltech).

will provide a different perspective of the great planet and the very real possibility of seeing something more than can be seen in the conventional view.

Fortunately, because of the success of the *Voyager 2* spacecraft mission, we were able to finally have a better sense of the appearance of Neptune. The spacecraft was able to obtain some remarkable and enduring images on its fly-by in 1989, which seem no less spectacular or beautiful today. Apart from vague cloud belts, we were able to observe some spectacular storms, not unlike Jupiter's Great Red Spot, although, once again, the colors remain in the blue-green spectrum. Vivid colors on gas giants are reserved almost exclusively for Jupiter, and to a much lesser degree, Saturn, at least in this solar system! (Fig. 12.4).

Quick Project: Viewing Uranus and Neptune

Time Required: 5 min per Planet

Catching more than a casual glimpse of Uranus and Neptune is more than worthwhile. Individually, their striking discs and haunting blue/green colors will justify the effort, as they readily appear much more than mere star-like points in the sky. Check positions on relevant sites or charts. They are not hard to find and only very slowly drift across the heavens. Be sure to wait for decent viewing conditions, just as with the other planets. This is because the diminutive disc sizes, and certainly anything unexpected, will otherwise prove difficult to resolve. You will need high powers.

It is worth stressing that it would be unrealistic to suggest that you will see for yourself, or be able to image by any method, views even approximating what is presented here! However, Uranus's and Neptune's place in the classical heavens make certain demands upon us as observers to visit them in any way we can. Considering the humble observing methods our ancestors used, we have capabilities beyond their wildest dreams and can thrill to live views such as they could never imagine; all they saw were star-like points in the sky. We also peer into our telescopes equipped with insights they never knew or suspected.

Pluto and Plutinos

It seems quite possible that much of the greater solar system remains as yet still undetected. Although the largest members of it have been well known to all of us since the earliest of times, in modern times the scientific community was temporarily lulled into complacency, believing there was little more to be discovered. Professional astronomers had even turned their collective gaze away from the solar system, and it was only with the advent of Space Age days that it has gradually come back into focus. So it has come full circle, with some remarkable and detailed studies and discoveries finally made possible. We live in a time when the classically known solar system may soon be joined by a large contingent of newcomers to the fold, at least planet-like objects in the remotest of solar orbits. Pluto has turned out to be anything but alone in the far reaches of the Sun's empire, or even in meager company, and it is entirely conceivable that there are an almost unlimited number of similar, relatively large objects orbiting the Sun. Many of them possibly could be much larger than the formerly named ninth planet.

The simple vexing problem of defining what a planet is and what it is not had long loomed over astronomers, even before they tried to rethink their newly expanding knowledge of the grander solar domain. At one time the benchmark for defining what was termed a planet was whether, during the formation of the Sun, matter in the original Oort Cloud had condensed through its own internal gravity into an independent sphere, locked in solar orbit as a complete entity from the start. Large objects such as comets were naturally discounted, since they had come into existence by gradually accumulating bits and pieces of leftover icy "rubble," stuck together only loosely by their own tiny gravity. And no structure other than one spherical in form was considered a candidate, because the possibility of its totality having condensed from the cloud in this way would have been eliminated. Therefore asteroids, no matter how large, similarly could never be counted as planets. But the age-old rules of simple definitions changed as more difficult to categorize worlds were discovered.

The topic of defining what constituted planet status returned once more to astronomical discourse only quite recently with Pluto's ever-controversial standing once again being challenged. The rancorous debate that followed served once again to focus scientists' attention back onto the question of the makeup of the solar system, and how to designate all of those objects making up the Sun's domain. It was finally voted on by a group of astronomers attending the IAU that in order for a potential candidate to be considered a legitimate planet in its own right, aside from orbiting the Sun and having sufficient mass to cause it to take a virtually spherical shape, it must, most importantly, *clear* the orbits of other planets. Because Pluto's orbit crosses Neptune's at times, this finally rang the death knell on its longtime planetary status. Scientists now prefer to categorize it as a "dwarf planet" or "planetoid." It has also been termed a "Plutino" (at least something was established in its honor), sharing this status with some other notable objects in solar orbit influenced by the gravitational forces of Neptune's vicinity. It does deserve at least some special category, since Pluto is clearly different from all the other long-known large asteroids lying between Mars and Jupiter, which clearly belong to a unique subgroup. Its orbital period, at 249.9 years, is dramatically greater than these asteroids, which typically have orbital periods of much less than 10 years. Although the scientific community seems set on its present planetary determination at this time, nevertheless poor, humiliated Pluto is again finding itself the subject of some future wrangling as to its status. It is not over yet. What would Clyde Tombaugh say!?

Nevertheless, Pluto's significance in astronomical history remains, and it will always carry a special status of its own. Because its predicted mass and final dimensions (1,500-miles diameter) turned out to be so much less than that had been anticipated, together with the uncomfortable facts about its orbit, the controversy surrounding it and on how it formed has always existed. Is it indeed, merely a Kuiper Belt object (so-named for the mass of icy bodies orbiting beyond Neptune), something perhaps formed out of the solar system's icy refuse? The recent discovery of another remarkably similar sized planet-like object early in this century (Sedna), and now yet one more (previously known "romantically" as 2003 UB313, now named Eris), together have done much to firmly establish the new solar system designation of dwarf planet. However, these are both more remote even than far-flung Pluto. And, it turns out, Eris is *larger* than Pluto, and so far away it makes our most distant previous "ninth" planet look like a neighbor just across the street. Thus, we have a new added impetus and meaning to the exciting new search just gathering steam in the early twenty-first century. It is impossible to say at present whether we will soon be describing an entire and newly known system. It all depends on what these objects actually are, how many of them exist, and what becomes the accepted definition.

Pluto was not the first planet to fall from grace. Sharing its newfound demoted position of dwarf planet in the solar system is long-established Ceres, which at one time also shared official planetary status! However, Ceres eventually "became" one of the best-known large asteroids instead. Meanwhile, numerous other asteroid-like structures (most prominently Pallas, Juno, and Vesta) are now being talked about in almost the same dwarf planet breath, even though they are not spherical in shape and belong more properly to the Kuiper Belt.

What does all of this do for us? Actually, not a whole lot, since we cannot appreciate much about these objects anyway in any type of observation. Although many are directly visible, particularly with such aids as image intensifiers, there is no way to distinguish any of them from background stars, making the comparisons of photographic plates necessary in most instances in order to recognize what we are seeing. For most of us, aside from the thrill of actually knowing we are gazing on such a world, there is very little visual appeal at the eyepiece, so it will fall to those who enjoy the pursuit and identifying of such pinpoint objects to spend the most time with them. And time is what they will need, for surely Pluto and any of its nonplanetary companions are not objects for a quick session at the eyepiece. Here, the value of a serious search utilizing CCD and other sophisticated methods of imaging come into their own, but such activities fall as far outside the focus of this book as these objects lie. Regardless, for the amateur, it is highly likely that our new neighbors will ever be accorded quite the same status as the old, because they will remain very inaccessible and largely undetectable, let alone allowing their surfaces to be viewable.

Thus, it would seem that the dwarf planets will always be regarded as lying on the "wrong side" of the orbital tracks by amateur observers. Nevertheless, perhaps your curiosity and available time will converge with the opportunity to see Pluto, at least, for yourself one day. By following charts in astronomical periodicals, we can readily know where to look and at what time, because without precise knowledge of where to see it, or by examining changing photographic plates, there is no way that we would ever distinguish it from the background of stars. A more complete approach would be to plot Pluto's position yourself using one of the *Starry Night* computer programs (www.starrynights.com). Even so, Pluto remains a problematic subject to identify positively because it is such a faint dot against the background, indistinguishable from tiny stars. It is easiest to recognize when it lies in close proximity to a plotted bright star in the sky, so wait for such opportunities before you spend too much time struggling to be sure of exactly what you have seen. Only fastidious attention and lengthy observing sessions to detect subtle movements over several nights will guarantee that you have succeeded in identifying it. And because Pluto is only a magnitude 14 object, you will need 10–12 in. of aperture to see it clearly, although it may be glimpsed in slightly smaller sizes. It is here that an image intensifier, or perhaps a frame integrating video camera, will pay huge dividends for the live observer. Not only will smaller apertures now show it much more easily, but larger telescopes will have a splendid advantage in revealing it as a brighter and more readily identifiable point of light. However, that is all it is to us: a point of light, and nothing to illustrate here!

The Hubble Space Telescope finally allowed Pluto to appear as more than the mere pinpoint of light that has become its hallmark in the eyepiece, though the mottled images obtained from Hubble ultimately reveal very little to those of us thirsty to really get a sense of the nature of its surface. Space probes eventually will change our knowledge of this remote freezing world, as well as many others on the edge of the solar system. NASA's New Horizon0027;s spacecraft is due to fly by the planet in 2015, so unless Pluto's new lowly status should sabotage the plan, by rendering it "unimportant," until then all we can do is speculate on its appearance and ultimate nature. As other planet-like objects are discovered in forthcoming years, with more time on your hands you might find this to be a dramatic new area to explore, but you will need time. Again, ideally, an image intensifier coupled to a moderate-to-large aperture telescope would be ideal for this live viewing purpose. The larger asteroids provide another similar telescopic mission and display a comparable appearance to Pluto's in the eyepiece. Many are being discovered that have dimensions qualifying them as more than mere "rocks in space," and in many cases may be readily detectable, especially if their locations are known ahead of time. And here once again we have the dilemma of defining what exactly constitutes a true planet.

Visitors from the Far Reaches: Comets

Comet Halley (Fig. 12.5) remains the single most famous comet of them all; unfortunately, its last appearance was something of a disappointment for most observers, and spectacular views were only obtained from exceptional sites. However, as if to make up for this, in the Northern Hemisphere we were truly fortunate in recent years to have been visited by some stunning comets, the most famous (and infamous for the reaction it provoked among one isolated religious sect) being Hale-Bopp. Its brilliance in the sky was reminiscent of those drawings from many years ago of such celestial apparitions, and its shape left nothing to the imagination, even from such light-polluted locations as Los Angeles. Unfortunately, such bright comets are quite rare in an ordinary person's lifetime, and expecting to see such things routinely is just not realistic, regrettable as it is to say.

Comets provide better opportunities for the busy observer than most other sights belonging to the farthest reaches of the solar system. They are spectacular

Fig. 12.5. Comet Halley 1986. Imaged by W. Liller with the International Halley Watch (photo courtesy of NASA/NSSDC).

in the best apparitions, and, unlike the others, they occasionally make house calls in this local vicinity. The irregular orbits that comets occupy allow them to do so only at the extreme end of their immense solar loops. When they are in the neighborhood the brilliance of the most prominent examples is tempting fare for those of us armed with large binoculars, low-powered large scopes, and especially image intensifiers, since we will likely have no chance to track them as near full-time observers. However, we must remember that comets are only visible because of reflected light, and even more importantly, the light that comes from their tails is very diffuse in its reflection, unlike that of the solid bodies of planets and asteroids. Despite the spectral response of image intensifiers, you may have good results with some brighter comets. Meanwhile, we all wait for appearances of truly great comets in the skies, where we need no optical aid at all for them to be grand spectacles. These rare visits always result in considerable excitement, even among those with no interest in astronomy.

In late 2007, Comet Holmes, first observed in the nineteenth century, appeared once again in the skies. Although it had undergone a transformation once before, it obliged us again during its most recent apparition with a radical expansion of its nucleus, creating a "halo" around it to the degree that the tail was completely enveloped. This phenomenon is apparently due to the nucleus literally blowing its material far and wide in a form of a (solar) heat-induced cometary explosion, something certainly rare in comets of the solar system. It is likely that most of us will never see such an event again during our lifetimes. The comet's *halo* became so wide that it almost filled the lowest power field of view at 81×. In the image (Fig. 12.6) the bright point of its nucleus may be easily seen, together with the tail remnants (pointing toward the upper left), mostly enveloped by the *halo*; a slightly darker region ahead of the core may also be seen between it and the leading edge of the structure. The appearance of many tiny stellar points shining through the comet made the sight even more magical and striking. Luckily, this author happened to be observing at a desert site near to the time of the comet's best appearance and was able to catch some revealing images.

Comet hunting is an all-encompassing occupation for many amateurs, some of whom have gone on to become legendary names in the comet-hunting field (such as Peltier, Levy, or Evans), with multiple comets named after them. But you will need time, and readers of this book are unlikely to have this in any quantity! Unfortunately also for the amateur comet hunter is the advent of sophisticated sky surveys, which cover every corner of the sky and which are slowly robbing them of much of

Fig. 12.6. Comet Holmes, November 10, 2007. Digital camera image by the author 2007 (18-in. telescope and I3 Gen. 4 image intensifier/2-s exposure). (AC)

their role. Extremely detailed scans of the heavens still need to be read by human eyes, however, so until that, too, is made redundant, some enthusiasts will continue the chase via the old fashioned method. Soon it would seem, though, that the only reason to take part in the hunt will be for one's own enjoyment, based on known coordinates, with no chance remaining of making any significant discovery. This is becoming ever more the lot in general of the amateur astronomer, however, and we must learn to be satisfied with it. There is still nothing like seeing the wonders of the universe for oneself, even if we did not discover any of them.

For more information and images from NASA's missions to comets, you might go to http://stardust.jpl.nasa.gov, where you also will find many links to other sites.

Quick Project: Viewing a Bright Comet

Time Required: 5 min

Unfortunately, for this project you will need to wait for an apparition of a relatively bright comet to appear in our skies! While not exactly a common event, such visitors show up from time to time and always are a source of wonder, even to the most jaded observer. They are especially suited to our purposes, since they do not need to take a lot of time to enjoy.

Do not miss any opportunity, and keep your eyes on astronomy web sites for announcements of a new visitor to our skies, especially for examples that are purely telescopic. Low to low-moderate powers will usually be best suited to them, all the more when you are trying to discern a long tail. Obviously, the darker the skies, the better. Drawing, CCD video imaging, or digital image intensifier imaging are all easy and quick options with comets, and the rapid evolution of their tails, brilliance, and other characteristics invite frequent, short viewings.

Asteroids and Minor Planets

We can be sure that any object in the solar system that is not spherical in form is not a planet, or even a minor planet. Any remaining controversy in defining planetary status will always surround those objects that are spherical, but for one reason or another do not quite fall into the actual category of being a planet. Never mind whatever terminology is used to describe any of these other worlds, however; they are all true worlds and destinations in their own right, some more than others. To see them for ourselves, we must first realize that the challenges are as great or even greater than with Pluto, so unless you can allot the required time for the search, it is not likely to be a viable option for you. Visually, they are indistinguishable from faint stars, and certainly no impression of their shape or nature can be detected.

Although there are no spacecrafts that have yet visited any planetoid, we are fortunate to have visited more than one sizable asteroid. Perhaps the most celebrated of those we have experienced at close range is Eros, with which the NEAR Shoemaker mission was able to rendezvous in 2000. Eros, it is now clear, is no planetary-type body, since it resembles a giant flying potato (!), while having impact craters of a variety found commonly throughout the solar system. Perhaps more interesting still was the rendezvous with the asteroid Ida. Not only was NASA able to obtain an impressive enhanced color image (by the use of infrared imaging), but it also was revealed that this asteroid had a moon (Dactyl) of its own! So tiny and close, it seems remarkable that this little cosmic pebble could possibly be suspended in such a seemingly unlikely balancing act. The asteroids have been described as rubble left over from the creation of the solar system. Although this may be the correct analysis, the asteroids are very large rubble piles indeed!

If you are not inclined to spend any time in these more remote corners of the solar system, perhaps now you can approach some of your viewing options for these subjects with at the very least a little better perspective. The most significant ingredients for more enlightened time with them are the new discoveries made possible only in recent years; this will certainly cause you to see them differently and only increase your interest to view whatever you can of them.

CHAPTER THIRTEEN

Daytime Astronomy

Perhaps the very term, "daytime astronomy," sounds like a contradiction. However, it relates mostly to observing the Sun, our nearest star. While the solar system's mighty benefactor remains the most important object of study for daytime activities, believe it or not there are some other things you can do during the day as well. In some ways unconventional, they are, nevertheless, more than merely trivial pursuits and thus may deserve a little of your attention, too.

Observing the Sun

Solar observing is an area of astronomy that has enormous appeal to many people, because it is the only star we can observe at close range. Indeed, it is also the only one we will ever see with our own eyes as a disc, and it may be seen as such even without a telescope! (The fact that it is almost a million miles in diameter is lost on most lay people.) The Sun also allows us to take part in meaningful astronomy during daylight hours, such as we may have available on weekends or holidays.

Unfortunately, the great potential lying right before our eyes must be matched by a large dose of *caution*. Observing the Sun has always been extremely dangerous for the unaware, cavalier, or careless. There is literally no time to react once the Sun enters the field of view – even with very small telescopes – before permanent damage has been done to the eye. In fact, it seems amazing that we do not hear of more such injuries, considering how carelessly many people use binoculars during daytime hours. Oh yes, a pair of binoculars is *more* than enough to do this!

However, using a combination of caution and appropriate equipment, countless solar observers and enthusiasts take great delight in spending as much time as possible

A. Cooke, *Make Time for the Stars: Fitting Astronomy into Your Busy Life*,
DOI: 10.1007/978-0-387-89341-9_13, © Springer Science+Business Media, LLC 2009

with our nearest star. By taking advantage of the sophisticated gear now available in the marketplace, they have been able to put aside reservations they may have had as observers using conventional means and enjoy regular views of this unique spectacle. One such enthusiast is John Watson, formerly an astronomy editor, who generously offered to provide some insights for this book. His enlightening comments follow in this section of the chapter. Since his methods and experience typify a modern practical approach, his contribution to this writing is more than significant.

As a starting point, let us look at some safe, proven, and simple methods for effective viewing. Although quite traditional, they are a good place to start. It is easy to quickly carry out many different observing projects with the Sun, and therefore they are highly suited to our purpose.

Some steps were mentioned in chapter 9 to radically reduce the amount of heat and light entering your telescope. Its optics are liable to explode through rapid overheating. Previous generations of amateurs used unsilvered mirrors in their Newtonian telescopes, but this meant a complete change of optics for nighttime observing. Aside from the inconvenience, the expense of having two sets of optics was something that made it largely impractical. The safety of using full apertures, even with unsilvered mirrors, is still unconvincing. Fortunately, large apertures are unnecessary for outstanding solar viewing, and some of the best amateur images have been taken with what would be meager apertures by normal astronomical standards.

Even though you may read accounts of fairly substantial apertures being used for eyepiece projections of the Sun, if you only have a large or moderate-sized telescope at your disposal, it would be best to stop down the aperture to no more than 3 in. Eyepiece projection may be effectively carried out with quite limited apertures and are more than enough for the purpose, since ever-larger telescopes pose incrementally increasing problems of heat and light. The great solar telescope towers in professional observatories were designed to utilize sizable apertures without causing the overheating experienced when using conventional telescopes.

Remember that the area to be stopped down on a Newtonian should be off-axis, not the region directly beneath the diagonal. With the primary mirror figured for a specific focal length, it now has the opportunity to perform as does a telescope of the same focal length, but with no central obstruction. Optically, it now essentially equals the finest refractor!

Ideally, the telescope as a whole should be sheltered from the heating rays of the Sun, and thus some kind of housing will be quite helpful, even if just a garden shed with a large opening allowing the end of the telescope tube to point at the Sun. This is in addition to producing enough shade for decent projected image contrast.

Most types of telescopes will work with the projection method. The Newtonian conveniently allows an image to be thrown onto a wall, though not in line with the Sun because of the placement of the eyepiece at the side of the tube. The screen will need to be fairly large to accommodate the Sun's track across the sky. Because other telescope designs will point the eyepiece toward the ground, perhaps this will necessitate using a diagonal, which can also be used to direct the image away from the light source. It is also possible to fix a projecting screen, constructed from white card, to the telescope focuser by stout wire arms, cut to the chosen length. Even better would be some kind of sturdy lightproof box, with a translucent screen at the far end. For projection on a screen, try to obtain the finest and least shiny white surface possible; this will make the resolution of detail far more telling.

That best known of all solar phenomena, sunspots appear dark (albeit with a lighter surrounding ring called the penumbra) because of their relatively cool temperatures, which contrast with the much hotter surrounding temperatures. Because of the nature of their makeup, sunspots usually occur in pairs (magnetic energy connecting one pole to another) and may exist in larger groups, with one pair dominant. As they wane later in their lifespan it is not uncommon for one spot to outlive the other, and interesting changes may be observed at this stage, as the Sun's brighter surrounding regions encroach and begin to reclaim their former fiery domain.

Quick Project: Indirect Solar Viewing Using Projection

Time Required: 30 min

Block off your finder scope (!), and locate the Sun by roughly aligning it along its length; it should not be too difficult to pick up, especially if you note the shadow it casts. *Do not use your finder; keep it capped!* The smallest shadow on the ground means that you are close to being on target. Use low powers, again to dissipate potential heat, because increasing magnification will destroy your high-power eyepiece collection quite quickly! With very low power, you may nevertheless be surprised just how large an image of the Sun can be projected directly from the eyepiece. Some of the most old-fashioned eyepiece designs (such as Ramsden and Huygenian) work best because they do not have cemented multiple lens components. Any heat can damage or destroy these cements. Do not use your fancy Naglers!

With a white screen squared to the telescope, try various placements of the telescope relative to it, all within a few feet at most. Despite the low power, you will probably be amazed at just how much detail may be seen, assuming it is present. Look for sunspots, as well as bright faculae – likely precursors of sunspots – at the limbs. Very occasionally you might even spot a solar flare leaping from the Sun's surface, although eyepiece projection is not the best way to observe this or the aptly described "granulation" on the disc itself.

You can, of course, photograph or record video of the images. And, oh yes! The Sun does rotate on its axis; you can follow the nearly month long rotation by following specific sunspots, which are likely to last long enough for the purpose, even although they can evolve quite rapidly. Should you want to know which are the preceding and following solar limbs, simply allow the telescope to stand still with its drive shut down momentarily; you will be able to orient the Sun and determine the direction of rotation quite readily from east to west. If you have spent any time observing Mars, you probably already know that significant drift can be readily observed on the disc, despite it occurring over a nearly month long rotation; it sounds slow but is easily detectable in quite a short time.

Direct Solar Viewing

For this approach a very low transmission neutral density filter must be secured to the large aperture of the telescope, together with an IR filter of some kind at the eyepiece end. The purpose of the large low-transmission filter is to drastically reduce the amount of light entering the tube. The least expensive types are made from Mylar film that has had metallic coatings applied to both sides; the solar image will usually be blue in tint. More costly types are made out of glass; not only are they sturdier, but they also allow a more natural coloration of the Sun's disc. Once again, the purpose of both types of filter is to absorb most of the light entering, but still to allow the transmission of a safe amount of red wavelengths. In any event, be sure to hold such a filter up to the light to look for any defects before you use it in conjunction with your telescope. Significant amounts of unfiltered light that produce multiple, even pinhole-sized defects may be more than you are prepared to touch up with dark paint, which is the preferred method for many solar observers. Never use any filter that leaves you any doubt; looking at the Sun directly is not something you should take lightly, regardless of modern technology.

At the other end of the optical train, an IR or H-alpha filter will be needed. As the detail-revealing characteristics of these filters increase, so does the cost, and it is not unusual for such filters to cost shockingly large sums, perhaps as much as the telescope itself! (You will likely never read or hear any comments disparaging the use of such devices because of cost. Contrast this with the illogical and frustrating arguments against using image intensifiers!) Nevertheless, such systems allow not only a more natural image coloration, but reveal much more of the textured *granulation* of the solar surface, and especially reveal solar prominences and flares strikingly, as well as superbly resolved sunspot detail; however, because of the small scale of some of this refined detail, the use of high powers will be necessary, along with sufficient aperture to resolve it. Some filtering systems have additional optical characteristics to change the focal ratio, which all but eliminates Newtonian telescopes for the purpose of direct solar viewing; these systems are designed for Schmidt-Cassegrains. Coronado Instruments, now part of Meade Corporation, have long made superior direct viewing Sun telescopes, complete with sophisticated filtering of multiple specified light wavelengths. They are expensive, but equipment at this level has never been cheap.

Quick Project: Direct Viewing

Time Required: 5 min

Simple....just hook up your filters and observe! This type of observing is among the most time-effective of all. It will be immediately apparent whatever detail is present, and the better your solar viewing equipment the better will be the quality of detail. It is also quite simple to hook up a video camera instead of an eyepiece (such as an Astrovid 2000 or a StellaCam) into the focuser and produce dramatic real-time views and moving imagery of the Sun's fiery surface.

Observing the Sun

John Watson, a former Astronomy editor at Springer Books and a true enthusiast of solar observing, has provided the following insights. If the Sun is to be a major part of your astronomy, and your time is limited, his guidance may well prove indispensable.

It's easy to take the Sun for granted, astronomically speaking. And yet it makes for an interesting and ever-changing target for amateur astronomers, one that is perfect for those of us who are often working in the evenings, or too tired (from working!) to get out to look at the night sky.

There are few of us who can't snatch a half hour during the day to get out and do a little solar observing. When I'm in the office (I travel a lot), I sometimes even set up a solar scope outside the building at lunchtime. It makes an otherwise dull pit stop for food more interesting!

I use two small telescopes for solar observing, and often keep one or the other in the trunk of my car. I also have a pair of tiny Binomite solar binoculars, made by Coronado (Fig. 13.1). They are very useful for a quick look at the Sun to check for sunspots. All but the smallest spots are visible, and it's a useful indication that it might be a good idea to set up the ETX or the PST. Holding the Binomites steady is the main problem. The design obviously had to be a compromise between enough magnification and the ability to hold the image steady. Sitting down helps, or leaning up against a wall. To use them,

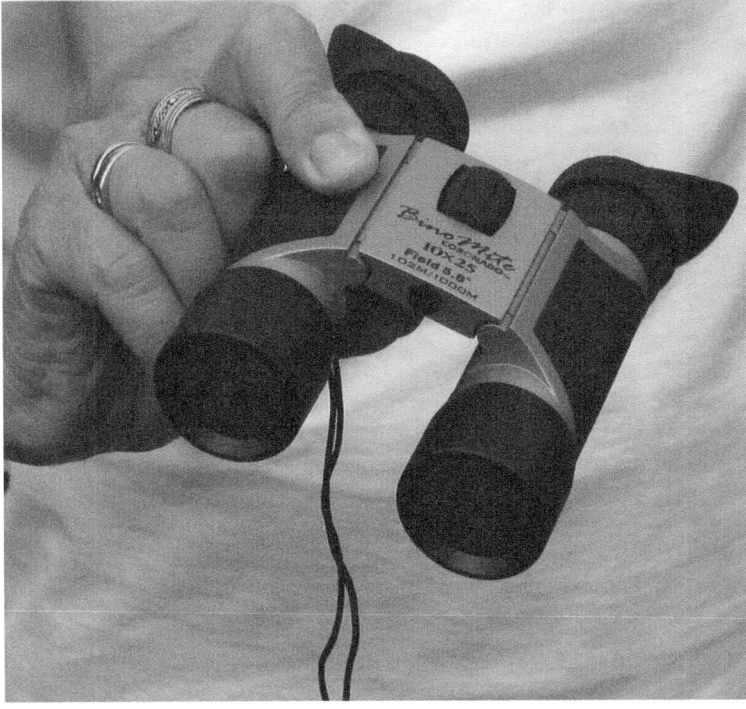

Fig. 13.1. Binomite binoculars. (JW)

aim *below* the Sun and then swing them upwards to make it easy to locate the Sun in what seems to be a black sky. When you've finished observing, swing them back down again before you take them away from your eyes, to avoid getting dazzled.

Meade ETX-90

The simplest model of the Meade ETX-90 is ideal for solar observing. It has an R.A. tracking drive, but no computer control. I bought it new quite a few years ago, but these days they're available used for very little money, as all the new models are now based around Meade's AutoStar system.

A good full-aperture solar filter is essential – I use an Orion – but that's all the extras, apart from a cap to block the front of the viewfinder. I use the top of an old 35-mm film can (remember them?!). It's important to put the cap on the *front* of the finder, not the back – otherwise the Sun's heat will instantly evaporate the cross hairs. It's better and safer to remove the finder altogether, if you use the telescope mostly for solar observing.

Set-up literally takes only five minutes.

Assemble the scope by attaching the three legs (or, if it's an AutoStar model, just stand it up). Place it on the ground or on a solid table if that's more convenient. The important thing to realize is that polar alignment doesn't matter much. If it's roughly right, then that's good enough. I keep a pocket compass with the eyepieces, and just line the scope up with the north pointer by eye.

It's as easy with the AutoStar models. When it selects an alignment star (which, of course, is invisible in the daytime), tell it okay, that's centered. It will be near enough for observing – and even imaging – the Sun.

The usual safety warnings for observing the Sun apply, of course. Check and double check that viewfinders are covered, and that the full-aperture solar filter is firmly in place over the front of the telescope.

Aim at the Sun by using the "smallest shadow" method (Fig. 13.2).

Swivel the telescope around until the shadow of the tube on the ground is at its smallest, meaning the tube is aimed at the Sun. Easy.

As I write this (Summer 2008), the sunspot cycle is at its minimum, and for weeks the Sun has looked like a cue ball, with no surface action at all. The good news is that for the next 5 years or more, the Sun will get more and more interesting!

If you're *really* in a hurry, you can check out whether or not it's worth setting the scope up at all by looking at NASA's SOHO (Solar and Heliospheric Observatory) on their web site at http://soho.nascom.nasa.gov/Sunspots/.

Coronado PST

Many years ago I saw my first solar prominence. It was at the home of Henry Hatfield, a noted U.K. amateur observer and instrument maker. He had built a spectrohelioscope in his garage. An enormously long-focus lens (the dimensions of the garage were designed around it) looked at the Sun via a heliostat, a plane equatorially mounted mirror up on the roof. At the viewing end he had made an oscillating-slit spectrohelioscope. It's too

Fig. 13.2. Aligning the Meade ETX-90 with the Sun. (JW)

complex to explain in detail here, but it basically uses a high-dispersion prism to look at one solar wavelength (e.g., H-alpha) and a vibrating mechanical system to scan the Sun's limb. When he built it, it was an amazing example of what an amateur could do with, basically, 1940's technology.

It's a lot easier now. The Coronado PST (Personal Solar Telescope) uses a "Fabry-Perot etalon" H-alpha filter system, with a bandpass of less than an Angstrom (0.1 nanometer, or 1×10^{-10} meters. It is thermally very stable. So it was that my new PST provided me with my *second* ever look at a solar prominence, through an instrument that conservatively cost about one twentieth as much as that wonderful old spectrohelioscope.

Observing the Sun in H-alpha is now an ideal activity for the hard-pressed amateur astronomer.

I began by organizing an equatorial mounting for the PST. I got hold of the cheapest equatorially mounted telescope I could find: the telescope was awful and I threw it away, but the tripod and mounting are quite good enough. My local motorbike shop made me a bracket to attach the PST to the mount. Manual tracking in R.A. is more than adequate for observing the Sun with a PST – easy to use and inexpensive.

Setting up the PST is again very simple. Set the mounting to your latitude and use a compass to get an approximate north. The PST has got a clever little solar viewfinder built in, for pointing the scope at the Sun without fuss or danger.

To check whether or not there is anything happening on the Sun in H-alpha before you get the PST out, see the Global H-Alpha Patrol Network image on http://www.spacew.com/Sunnow/. It's updated every minute!

Fig. 13.3. Coronado PST, universal digiscoping adapter, and camera. (JW)

Imaging on the Run

You can image the Sun in H-alpha with the PST and digital camera.

I use a Universal Digiscoping Adapter to attach my little Canon Powershot SD450 to the eyepiece tube. The principle is the same with any other attachment bracket (several are available) and a compact digicam. The camera and bracket together weigh just over a pound (0.5 kg), and the mounting's balance isn't affected enough to require counterweights or anything like that (Fig. 13.3).

Set up the PST to look at the Sun's limb, and when you have a clear view swing the camera into position. Here's how the camera's screen looked - Fig. 13.4.

Fig. 13.4. Camera screen with solar image. (JW)

The autofocus should work fine, provided the camera can "see" the edge of the Sun. Use the self-timer to allow vibrations in the mounting to die down, and make your image. I found that zooming in was a waste of time. It's better to enlarge part of the image when you process it on your PC.

At this magnification it isn't necessary to track the Sun. The exposure time of the shot below was 1/90 sec. with the camera set to ISO 125 (Fig. 13.5).

This image was processed using Paint Shop Pro 8, to enhance contrast at the limb. I'm not citing it as particularly special image, only that it was taken 'on the run' one afternoon when the Sun suddenly came out. It's typical of what to expect from this set-up first time around. The image is actually overexposed: monochromatic light doesn't seem to work properly with the automatic exposure control of most cameras. If you've time, take a range of shots at different, manually selected, exposures until you come up with the one that works best. Then write it down.

The stunning photograph by Massimi Lorenzo on Coronado's web site at http://www. coronadofilters.com/products_pst.html shows what can really be done with a Coronado solar telescope (although not a PST).

And probably not during a sandwich break.

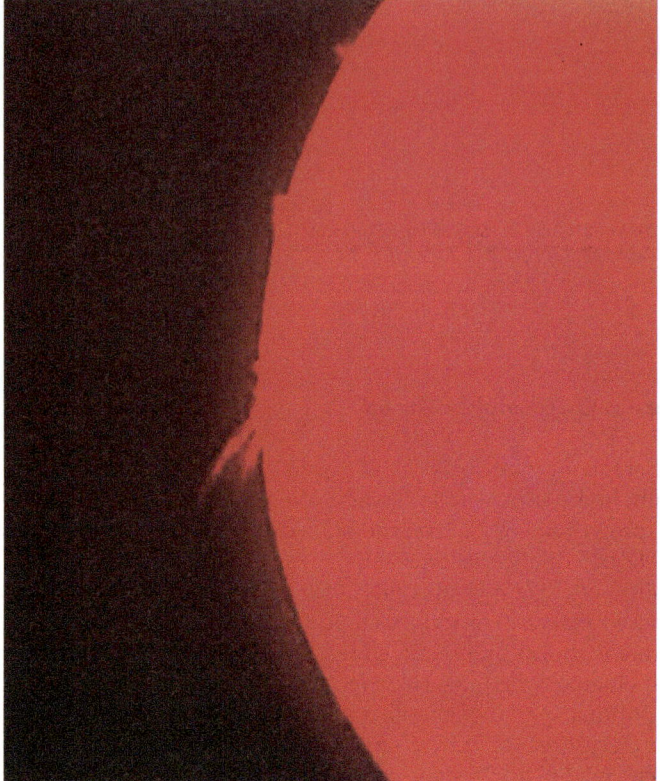

Fig. 13.5. Solar limb. (JW)

Viewing the Planets During Daylight Hours

Perhaps surprisingly, the brighter planets also offer us some possibilities for viewing during the day. Almost needless to say, this is seldom tried or even the subject of much commentary. Perhaps there will be some new technologies that will give it even more potential down the road. In the meantime, there is no doubt that seeing other worlds at times when you would least expect to do so has a special appeal all of its own!

You need to bear in mind that daytime viewing also has the effect of making objects appear paler, because of the diffusing affect of the sunlit sky. You only have to casually think how the Moon appears in the daytime to be aware of this. Different filters may be helpful to try for such far-from-conventional viewing, including polarizing and light pollution filters; remember, anything goes if it produces results, so do not be afraid to experiment! There has been surprisingly little commentary from the amateur community about this form of astronomy, so there are very few absolutes established at this stage.

The inner planets, Venus and Mercury, will still prove to be challenging objects, except now you will have a chance to observe them earlier when they are more favorably placed (higher) in the sky. However, never forget the close proximity of the Sun to them! You will need to locate these planets by ways other than using your finder scope. However, because their brilliance will be much reduced, they may show you some detail. With far less swirling air to look through than when they are low in the sky during the early evening or morning, this has to rate as a very big plus, in addition to the more stable air temperatures. It would therefore seem that trying your luck in the daytime has a double bonus. With a violet filter, Venus's elusive cloud markings may oblige you more easily in these conditions. Just remember the atmosphere needs to be just as conducive to astronomical viewing as it does at night!

Less successfully viewed in comparison to the nighttime will be Jupiter and Mars, which, under normal observing, are such a great resource of color and contrast. However, even during daylight hours, it is not out of the question to see cloud belts on Jupiter, even if they are rendered essentially colorless. Just how successful might you be? In the previously mentioned volumes, *Amateur Telescope Making*, there is a reference to a small refractor revealing cloud belts on Jupiter during the day. This was written at a time (1930s) when such radical forms of observing must have been virtually unknown; just finding the planet without digital circles was a feat in itself!

Mars is unlikely to be near enough to opposition to see with much frequency during daylight hours (because of rising in the sky mostly after dark), but this should not be taken to mean that it is not worth trying when circumstances are right. In daytime, one would normally expect all but the most obvious detail, such as the polar caps, to be washed out in the view. The prospects do seem tantalizing, though, and maybe some degree of ingenuity and the use of an appropriate or unconventional filter might just surprise the most pessimistic observer. Mars is so bright at opposition that its disc will certainly be visible in your telescope's field of view; who knows, maybe you will be lucky enough to see the Syrtis Major. A low expectation for Saturn might turn out to be unfounded. With sufficient aperture, it is possible to pick it up during the day, but you will not be likely to see any kind of detail on it at all at these times. However, just seeing the ringed planet during these unlikely hours will bring much satisfaction, especially to those who have never before looked through a telescope.

Quick Project: Viewing the Brighter Planets During the Day

Time Required: 10 min

During the day, and it cannot be stressed enough or too often, *you must take enormous care not to snare the sun during your wanderings around the sky*. It is all too easy to do this with Venus, and even more so with Mercury (which follows the Sun very closely), so be sure to use your digital setting circles *exclusively* for the purpose of locating these objects. Even then, take special precautions to make sure that the telescope is well away from the Sun's disc. Do not take any risks; it is just not worth it. And finally, be *very* careful as you approach the eyepiece, just as a final precaution. Naturally, you will need to have your scope set up and star aligned the night before. Use any and every filter you have to look for maximizing whatever you can see.

Just glimpsing any of these planets at such unlikely times is rewarding, but this is soon followed by the challenge of discerning detail in the discs themselves; because you will not likely see great amounts of detail, your observing session will not require a great deal of time, so what could be better?

Viewing the Moon during the day may not hold much fascination for you. Just because it is so easy to spot during daylight hours takes much of the challenge away! With the wealth of detail so readily visible at night, along with its great availability (unlike the planets, with their specific oppositions and viewing times), trying to carry out any type of lunar observing during the day does not seem to make much sense. If you do try, the washed out view compared with what you might see just a few hours later will disappoint you.

Observing During Twilight and Early Morning

As an extension of daylight viewing, we should also consider those many times when it is neither strictly day or night. Early evening, just after sunset, can be a wonderful time for many of the brighter objects in the sky, as long as temperature falloff is not a factor. Unfortunately, much of the time it is. In some instances these conditions can bring about some of the best observations of all, because there is sufficient darkness to allow nearly maximum contrast, and the discs of the brighter planets will not overwhelm your eye with the stark brilliance that has become their hallmark. This sometimes allows the detail to be pushed into the visible foreground. These times are also extremely good times for lunar observations, with no reason to wait for full darkness for outstanding results. Even a fairly bright background sky is no impediment to good lunar views, as long as thermal issues do not intrude.

Other Daytime Prospects

Quick Project: Seeing Stars

Time Required: 5 min

Believe it or not, some of the brighter stars, which we will talk about in the next section, may be seen quite readily during broad daylight. Of those that are visible, the red stars seem most striking in these conditions, and they have qualities in the field of view quite unlike their later appearance – altogether purer and of clearer color, without any of the dazzling or overwhelming brilliance that creates an entirely different impression at night. It is in many ways akin to seeing some of the grander sights at night, just for the beauty of their colors and brilliance in an unlikely sky.

Try observing Betelguese or Regulus, for example, and do not let the blue-white of other brilliant stars such as Sirius stop you from trying to view them, too. Again, the same precautions apply, including the need for your setting circles to be aligned ahead of time. At times there will be other objects on view as well. Perhaps many of the brighter comets could be picked up in this way, and may well provide some very interesting viewing. One observer was able to see Comet McNaught during daylight hours! The sky is the limit!

Section IV
Deep Space

CHAPTER FOURTEEN

Viewing Deep Space Objects

Deep space is our ultimate and grandest destination. It can also be the biggest challenge and disappointment when inappropriate equipment, circumstances, or approaches are used. Although shortage of time and opportunity may limit successful observations to just a few objects during any given session, this will prove more than enough! However, because most of these subjects are so inconceivably distant and faint, they seem to hide at the very threshold of visibility. Indeed, deep space destinations will always keep some of their characteristics shrouded in a veil of secrecy, whatever we do. We will always be limited to knowing only a tiny part of the total reality of any object; even the term "object" is strangely inappropriate to describe such vast and complex entities. It would be comparable to an extraterrestrial life form across the universe describing the entire Milky Way Galaxy, and all that it encompasses, as an *object*!

The visual vagueness of these destinations (a better choice of words), and the time it takes to chase out each one is why success in deep space is usually an issue. The good news is that the difficulties of succeeding need not be insurmountable these days, if we take proper advantage of what is available. It is possible to carry out some very worthwhile observing, even if we only can do it in short bursts, and from less than ideal locations. Nevertheless, it would be foolhardy to suggest that you should not seize each and every opportunity to relocate to dark sky country, even if doing so must be a rarity. Assuming that virtually everybody has the chance at least once in a while to relocate to a favorable viewing place, there is nothing like it to reinforce all your other viewing.

If you find yourself largely city-bound, or just too busy most of the time to get to an optimal location, you can ill afford to waste the opportunity when you finally can relocate to one. Proper planning of your time with the telescope is yet another important factor in how much you make of it. There is a clear relationship between

A. Cooke, *Make Time for the Stars: Fitting Astronomy into Your Busy Life,*
DOI: 10.1007/978-0-387-89341-9_14, © Springer Science+Business Media, LLC 2009

suburban viewing, dark sky viewing, and the topic of this present book, because having the capability for worthwhile astronomy from home is presumably something you *need* to do, and taking it away from home is something you will *want* to do. Today, the best in observational equipment and accessories make this a much more realistic proposition and has added greatly to our capabilities and efficiency.

Since the advent of modern CCD imaging in all of its guises, compelling reasons to actually view these sights live through the telescope, let alone to find encouragement to do so, have become a rarity. "Deep space" has come to mean something indirect, quite labor-intensive, and usually involving added time away from the telescope, too. Unless you were to know better, from all that you may have heard and read it might well appear that there is little point in chasing such cosmic ghosts with your own eyes. However, with just a little viewing skill, and a little better than minimal conventional equipment, you can see far more than you might have been led to expect; from dark surroundings, you will be truly amazed. With ever better tools and know-how, the amazement will continue to grow.

Although it is certainly true that nothing we will ever see through a telescope will be able to compete with the dynamic range, or the dazzling and brilliant colors of the images we see propagated so much today, it is also fair to say that at least what we may see with our own eyes will be closer to the reality of the universe. This is because even if we *were* able to relocate to advantageous viewing positions in space, most of the objects there would *still* look pale and diffuse! In reality, as briefly commented in Chap. 9, the universe is a much more subtly colored place than we might believe.

Consider, for example, our own location right inside the spiral arms of a truly great galaxy, the Milky Way. Even under the darkest sky, with the naked eye what we see of it is still quite faint and bland, despite the fact that it is comprised of hundreds of billions of stars of all colors and levels of brightness often far larger than our own Sun, and yet we are right in the midst of it! To us here on Earth the combined stellar colors are largely perceived monochromatically, as one color – white. When we look carefully and isolate individual stars we can see subtle colorations, of course, or even many colors when examining them through a telescope. But the effect of looking out into space at the great vault of the heavens still remains largely one of a monochromatic entity. Even the huge illuminated nebulae, while radiating great bands across the light spectrum, would still appear faint because of their tenuous nature, and essentially colorless to the eye, regardless; in fact, were we situated up close, they would probably be invisible!

It is not a completely blank canvas, though. As in the case of individual stars, some colors of the grander clusters are very apparent telescopically in their makeup (blues and yellows especially), and certainly some of the brighter nebulae show color, or striking hints of it. However, we must examine them with sufficient magnification and aperture to draw attention to these qualities in ways meaningful to our eyes; none of it is apparent during a naked eye scan of the Milky Way.

There is, however, one other part of the equation that also removes color from view. When we are studying deep space objects in dark conditions our dark-adapted eyes become greatly desensitized to color; you may notice this effect after a few hours of stargazing even from far-from-dark suburban home locations in the early hours when many lights have been turned out. You will notice that under these conditions we become most sensitive to green wavelengths, which explains the logic behind the selection of monochromatic green phosphor screens in image intensifier tubes. Colors that were vivid become bland; green parts of the spectrum seem more prominent, and even the irritating color of sodium vapor street lights no longer seems to be such a predominantly orange hue. This highly sensitive part of our vision (green) is made dominant in order to emphasize any scrap of light that lands on our retinas. So, it is when our eyes have become most sensitive to low levels of light that they are least able to detect color! A conundrum indeed.

The brain seems to disregard even the green spectral sensitivity, and most deep space objects simply will appear just white or grayish-white to the dark-adapted eye, if they did not before! Vivid hues are likely only to be revealed in long-exposure imaging, and typically involve substantial later processing to reveal the vivid colors we have now become accustomed to seeing in illustrations. However, considering the fantastic scenes that it is possible to see live, the mostly monochromatic viewing we experience at the eyepiece is a small downside; in fact, it is much truer to life! If you are observing from home, you might try viewing in the early hours of the morning. Quite aside from the advantages of having your telescope in fine thermal equilibrium (assuming you have set it up in the early evening), many city, commercial, and neighborhood lights will finally have been turned off. The difference can be remarkable, and you can accomplish some far better viewing than you might think possible from populated places. You just have to make yourself get up from a good night's sleep and step into the cold night air!

Despite the absence of color in most live deep space viewing, there is still absolutely nothing to prepare you adequately for what is possible to see in the eyepiece with just a little care. This is only appreciated all the more as you immerse yourself deeper and teach your eyes to "see." The best part is that no image, no matter how well executed or spectacular, will ever become a substitute for the real thing, which is also far more immediate and direct. This basic fact seems to have been overlooked by many enthusiasts today, as live viewing is being relegated faster into obscurity, only to be replaced by something quite clinical, indirect, technical, and complex by comparison. Make no mistake: looking directly through a telescope represents the essence of traditional amateur astronomy, "outdated" though many "new astronomers" would tell you that it is. The best part is that it can be done in short observing sessions that begin and end at the eyepiece!

For reference materials, two sets of books are highly recommended: *Celestial Handbook*, by Robert Burnham, still perhaps the foremost guide to deep space observing. Its three volumes make great reading when you are far away from the telescope, too, so you can enjoy your interest at other times with some of the most inspirational words and sentiments you are ever likely to read. Then there is Kepple's and Sanner's *Night Observer's Guide*, Willman-Bell, Inc., which also has its special respective place as the ultimate field guide, with untiring descriptions and countless photographs of every deep space object you can imagine. Both of these sets of books were put together by true visual astronomers. Despite the fact that they were written by the exact opposite breed to the enthusiast trying to sandwich a little observing here and there, they are invaluable resources. The combination of both sets (five volumes in all) should provide enough information for the avid night sky junkie indefinitely. As an observer you could not wish for more, except perhaps, the complete Southern Hemisphere coverage in the *Night Observer's Guide* (missing in those volumes), if you happen to live there!

Near Deep Space

Although anything in deep space is remote by any standards, many of the grander sights for the visual observer belong to our own galaxy, and cosmically speaking are local objects. We can count among these sights star clusters, nebulae, planetary nebulae, and individual stars (often doubles or multiples). You will find that it is these that will provide the majority of easily accessed spectacular sights for those with little time to spare or confined mostly to suburban home locations. Because of the relative proximity of these objects they provide viewing qualities that you will usually not find outside the Milky Way, with some notable exceptions, of course. Unfortunately, CCD imaging may have given you unrealistic expectations that everything in the cosmos is just waiting for you to view live at the eyepiece, and in brilliant colors at that! Actually, what you will find instead is somehow all the more wonderful, once you have fully developed how to *see*.

Star Clusters

As some of the most immediately spectacular and readily viewable deep space objects, star clusters are ideal material for us! No novice could fail to be enthralled by the sight of a great globular cluster, although many fine examples exist among the more sparsely populated open clusters. However, most of the latter will prove low-power objects, because, as relatively loose collections of stars, they cover wide fields of view and are consequently best seen with very low powers, or even binoculars. The makeup of star clusters is often well suited to live viewing, even when conditions are much less than ideal, including viewing in polluted or bright skies. They may come closer than you might think to their expected appearance when seen under highly favorable viewing conditions and are easy to find and enjoy when time is of the essence. They will accommodate many short sessions, always providing varied and enthralling viewing. (Refer again to Chap. 3, Fig. 3.1 – Hercules Cluster M13, taken in suburban conditions, and, aside from the green hue of the image intensifier utilized for easy imaging, this picture is quite representative of a conventional view at that location with just a moderate aperture.)

Open Clusters

Open star clusters, such as the Pleiades and the Beehive, are well known even to the most casually aware skyward-gazing individual. Most owners of a pair of binoculars will have looked at some of the brighter open clusters; in fact, they have actually stumbled onto one of the best ways to view them, and most are best suited to binocular or other low-power wide field scopes (i.e., that seldom heard term these days, "richest field telescopes"). As such, they are often something of a disappointment through moderate and larger telescopes, where the spread-out stellar populations of these clusters prove to be just too far apart to fit in the field of view. There are, of course, some notable exceptions, such as the dense and interestingly shaped Wild Duck Cluster M11, and the "Sword Handle" NGC869/884 double cluster in Perseus. Both of these sights, well populated by stars and by their distance compactly presented, look nothing less than spectacular through even quite large telescopes, rivaling many a globular cluster.

The stellar makeup of most clusters appears in varying colors, with different clusters revealing different things about their age and composition. (Blue stars are young, white stars youthful, yellow stars middle aged, and red stars are old.) Natural views are usually exquisite in appearance and offer some of the most satisfying viewing of all. It is regrettable that image intensifiers cannot show these colors, since everything appears in monochromatic green; never mind, they have other qualities for observers instead! Frequently you will see surrounding nebulosity in the newer clusters, which are the birthplaces of stars, condensing out of the cosmic fabric, such as that surrounding the exquisite pale blue stellar components of the Pleiades Cluster M45.

If your sky is dark enough, do not forget to spend the occasional session with the great star clouds of the Milky Way itself! They offer quick and ready access and considerable enjoyment for very little time spent. The Sagittarius Star Cloud M24 is one of the most celebrated examples. Although these Milky War regions are not star clusters in the normal sense, some of the densest parts of that great band in the sky make for some pretty impressive viewing. Many great dark nebulae will also be present in these scenes, such as the area surrounding the hub of the galaxy in the region of Sagittarius and also to the huge clouds of stars in the region of Cygnus. Here, you will need binoculars, a richest field telescope, or at least the widest field with low power, and fairly dark (but not necessarily darkest) skies, in order to view these destinations in spectacular fashion. You can accomplish even more with an image intensifier eyepiece attached to a nonmagnifying attachment, as previously described. However, these devices open up all kinds of possibilities, and you never know what you might see – even things apparently not in any catalog (believe it or not)!

Globular Clusters

Little competes with the telescopic glory of the brighter globular clusters. Because they are relatively dense, compact formations, they are ideally suited to the small fields of view of most telescopes, as well as being readily observable and accessible in less than perfect skies. There is nothing diffuse about their appearance; either they can be seen or not. An entire session can be spent reveling in these spectacular and varied sights. Beyond the most readily obvious qualities that jump out at first glance, you will not find two alike. Many of the larger ones (at least in angular size) are well suited to moderately high powers without their outer stars spilling outside the field of view. All told, including the less celebrated examples, you can probably find at least fifty well worth your time.

When studying a globular cluster, you might note that before the eye connects with any existing visible dark lane, it might not seem to be there at all! However, once sighted, it will tend to stand out quite conspicuously, now hard to imagine that it initially seemed hidden! You may already know that the famed "Propeller Lanes" in M13, first reported by Lord Rosse during the 1800s, were once considered "lost" through much of the twentieth century. Perhaps the off-center position of these lanes in the cluster was responsible, since Rosse's descriptions and drawings were somewhat different to the way we know them now. However, amazingly, they are readily visible on some old photographs. With our new hindsight and a little indirect vision you will find that they never did disappear. Interestingly, you will not find any comments about their existence at the time these photographs and observations were made! Even after having seen them live countless times, they sometimes appear none too obvious at first – until once sighted, they suddenly dominate the view!

With globulars, even a half hour's viewing will allow you to see more than enough to be deeply satisfying. The less prominent examples are almost as spectacular as the better-known brighter ones if you have the equipment (translate: aperture!) to view them effectively. You will find great satisfaction in splitting the far distant globulars into at least some stellar components, something made quite easy with image intensifier eyepieces. (See the reference to M54 in Chap. 3, certainly a favorite example.) There are others, too, among them the famous "Intergalactic Wanderer" NGC 2419 in Lynx, which is apparently a freak of nature, since it does not seem to belong to our galaxy or even any other and lies at a distance substantially further from us than the Milky Way is wide. Its independent existence remains baffling, but it is not the only case in point. Another to scrutinize in favorable circumstances is NGC 7006 in Delphinius, another wayward and remote cluster. Without some means to boost the faint stellar points, either through intensification, frame integrating CCD video, or even great location, these intergalactic subjects are most likely to appear merely as fuzzy blobs, easily mistaken for elliptical galaxies. You will not hear even the owners of quite large amateur telescopes describing much stellar resolution with conventional viewing. Usually, at best, the term "granular" is what is used for their appearance in the eyepiece, although this is no reason not to visit them.

Taking the ultimate step, you may want to take the trouble to research and locate globular clusters in other galaxies, such as those in M31 in Andromeda or M33 in Triangulum. It is a little like seeing Pluto – not so how much you see as the knowledge of *what* you are seeing! However, as you might expect, you may not have the luxury of sufficient time to make this meaningful, because they are pretty research-intensive objects, although a few may be easy to locate.

Unfortunately for most Northern Hemisphere observers, the two grandest globular clusters happen to lie in the Southern Hemisphere (Omega Centauri and 47 Tucannae). From prime southern California locations, the biggest and baddest of them all – Omega Centauri – is worthy of an entire observing session. Dancing low on the horizon, it is nevertheless an awesome sight, its stars completely filling the field of view and beyond, with scarcely any dark space at all being visible in the field. Its low position barely detracts from its splendor; it is so bright and heavily populated with stars that one would think it *ideally* placed. Unlike any other globular, certainly the theory that this globular is the core of a captured dwarf galaxy seems plausible enough.

If you have any chance at all to see it with your own eyes, do not miss it, because it is in a different league to all of the rest; the sight is unforgettable. Second only to it is 47 Tucannae, which regrettably is not visible from the continental USA. It more than lives up to its reputation: grand and star studded, it is a treat that unfortunately most Northern Hemisphere residents will seldom see.

Quite aside from these two grand spectacles, residents of the southern skies are also surely blessed with some additional splendors, which we Northern types must forgo; they also see the majority of *our* best sights too! The great Southern spectacles include the Magellanic Clouds, the Tarantula Nebula, even the dazzlingly beautiful double-star Alpha Centauri (astonishing in the simplest eyepiece!), to say nothing of the superb overhead placement of the hub of the galaxy and all of its attendant wonders.

Quick Project: Touring Bright Clusters

Time Required: 5 min per Cluster

Star clusters, both open and globular, make ideal subjects for some very effective, easily carried out astronomy. Binoculars (the larger the better) will often provide the most ready means of touring many of the open clusters, leaving telescopes to reveal globulars to best effect. When time is short, it will probably be best spent with the brightest examples, a few of which are listed later, showing a variety of qualities. They are all telescopic objects. (See Chap. 17 for numerous additional examples.)

M11 – The "Wild Duck" cluster. Brilliant, with a leading line of stars shaped like an arrowhead (conjuring up the image of a flight of wild ducks), this is certainly one of the finest such sights in the sky.

NGC869/884 – The "Sword Handle" double-open cluster, with the stars situated close to each other, features many multicolored, relatively dense stellar populations.

M2 – Exquisitely beautiful and brilliant globular, M2 is quite compact and just the right size in the field of view under moderate power. A magnificent white snowball, the fairly evenly matched sizes of its stellar members, none of which dominate the cluster, make it a favorite.

M5 – A perfect sized globular in the field, bright and beautiful, this object shows many dark lanes crossing it at all angles, most apparent with image intensification.

NGC 5139 – Omega Centauri. There are more stars visible in all directions than can possibly fit in the field of view, seemingly at any power! If you are fortunate to be in a sufficiently southerly location, there is nothing else comparable. Indeed there is not anything even close!

M22 – Quite a contrast is this grand and relatively open globular cluster, which features stars of many levels of brightness, covering an area far larger than most but not too much for low to moderate powers. This is one of the most spectacular and magnificent sights in the sky.

M4 – No cluster shows such amazing, almost bizarre, loops and chains of stars as does globular M4, but it only reveals its very best traits under truly dark skies, where the darkness allows the many voids to be filled in with stars.

M13 – This is the legendary granddaddy of Northern Hemisphere globulars, with surprising spiral star arms reaching out all around giving it the appearance of a cosmic spider. Considered the finest Northern Hemisphere globular, it also contains the famous dark "Propeller Lanes."

M3 – One of the finest globulars, compact and bright, M3 has a lopsided placement of its core; it is also full of dark lanes of obscuring matter.

Diffuse Nebulae

Diffuse nebulae also rate as among the most impressive and readily viewable objects in our "local" deep sky. They present us with multiple viewing opportunities and almost limitless variety. Many examples perform surprisingly well from suburban locations, so for those observers looking to sandwich a little viewing into a moment here and there, they are well suited to the task. However, there are distinct types of nebula, often significantly different from each other in character and origin, some more readily observed than others. Many are intermixed.

Emission Nebulae

Good and immediate results at the eyepiece are especially likely with those nebulae falling into the emission "camp." Not only do these often show quite wonderful detail, but they also respond so stunningly to any form of telescopic viewing that they surely rate among the best deep space subjects of all. This is not to say that other varieties (reflection nebulae) are disappointing, but only that those are less consistent with results and offer less latitude in viewing approaches. Try narrowband filters on emission subjects, even from dark sky sites; you might be pleasantly surprised. Emission nebulae also respond wonderfully to image intensifiers and frame CCD integrated video. In many instances, CCD video cameras mimic the response of image intensifiers quite well, although without quite the immediacy or finesse. Either way, suddenly what was average becomes dominant in the view, and even partial emission nebulae (those mixed with reflection gases as well) usually provide incredible sights. The views are frequently comparable to time exposures, except now they have the impact and presence of live viewing. There is scarcely a brighter emission nebula that does not put on some kind of show, regardless of the method of observing.

Meanwhile, among subtler emission nebulae worth some effort is the Veil Nebula complex NGC 6960/6992 in Cygnus; with a narrowband filter you might try your luck with it from a city location. If it is to be a successful target, you will know it right away. Yet, it really comes into its own from a dark location, further revealing structure and twisted filaments from this old wreckage of an exploding star. Unlike this dying remnant of a past disaster, most of the larger emission nebulae are cradles of hot young stars, in various stages of becoming of fully fledged suns, whose growing nuclear infernos excite the gas clouds into bright luminescence. Here, you may witness creation firsthand. Skillful viewing frequently will allow you to count seemingly innumerable young stars, while enhanced viewing of any kind only makes this all the more spectacular and immediate.

Reflection Nebulae

Because reflection, emission, and dark nebulae are frequently seen as parts of larger whole nebulae, it is not always easy to find many examples that are exclusively of any one type. When the reflection component is clearly dominant in any nebula in particular, conventional viewing will often be found to produce the best results. From areas of light pollution many reflection nebulae are likely to be more problematic to see than many other *local* deep space objects, making them harder to justify excessive attention when your time is short. Image intensifiers most likely will be less successful with them than other means, since the wavelengths common to reflection nebulae are not where the intensifier's response is strongest. However, CCD video will often perform well on many subjects, including reflection nebulae. When time is short, and you are viewing from your home base, only the brighter subjects will usually be worth your while. However, from darker places, these particular celestial wonders are some of the most delicate and subtle formations you will find anywhere in the sky. A prime example of a pure reflection nebula would be the puffy thin veils surrounding the hot young blue stars of the Pleiades Cluster.

You will soon realize that under good viewing conditions, there is scarcely a nebula – of any type – that will fail to put on some kind of show with conventional viewing. However, from less good, typically suburban locations, an Orion Ultrablock narrowband filter proves quite useful for conventional viewing of most types of nebulae. These filters work across most of the spectrum most likely to be viewed. You may have other types of narrowband filter in your arsenal that work well on these subjects, too, and for pure visual enjoyment, there is little to touch the value of these accessories. At dark sky sites, some *broadband* filters may prove to be of value, too, although they may not be especially useful in suburban locations, their application being too general to help. From suburban locations, do not waste short observing sessions with the fainter reflection nebulae; they will be disappointing at best.

To see many of the legendary sights to maximum advantage you will need to relocate to dark skies. Savor these times if they can only be infrequent. Such spectacles as the famous Eagle Nebula M16 can reveal their true nature; truly astounding in the field of view, the great bird looms out of the clouds of gas and stars, although the eagle itself is the result of dark nebulous clouds in front of the illuminated portions. The great bird's wings are spread, with its talons ready to pounce, with what appears to be a glorious wake trailing far behind. However, you will need some developed viewing skills to make out all the details; you should not feel discouraged should you have neither an image intensifier nor a sophisticated integrating CCD video camera, such as the Astrovid StellaCam III.

Dark Nebulae

Dark, unlit nebulae are interspersed among most of the bright nebulae and star fields in general throughout the galaxy. Indeed, even the famous bright nebulae would probably fall into the same category were they not illuminated by reflection or ionization from embedded stars. So it is quite normal for dark voids to be present amid many of the bright nebulae, helping to give added dimension to their apparent folds and swirls. It is easy to confuse what we are seeing as being actual folds in the fabric of nebulae, when in fact they are more likely to be the result of random mixing of all types of nebula, and at varying distances in front of what is lit. Bear in mind though that in suburban locations all but the most dramatic of these nebulae may be rendered almost invisible. Therefore, they are best observed (or to put it more accurately, *not* observed and their light-blocking effect realized) under truly dark skies, because of the need for maximum contrast; probably you can count most of them out for brief encounters with the stars from your home site.

In regions of dark nebulae it seems as if the stars have just been snuffed out, but just the opposite is the cause! Sometimes, instead of immediate blocking, the stars just seem to fade out into the denser nebulae, creating an eerie awareness of the blackness and seeming infinity of deep space. Certain examples will be even more stunning when observed through enhancement devices.

Quick Project: Viewing and Comparing Diffuse Nebulae

Time Required: 5–10 min per Subject

What is most striking about viewing certain nebulae using different methods is the radically varying emphasis of different parts of the whole. This is most noticeable when emission and reflection types are generously mixed. When time is of the essence the most celebrated examples will seldom disappoint. You will never tire of all the detail, or the sheer brilliance and ease of viewing, regardless of the method of viewing. However, by all means at least try a narrowband filter! Again, as with so much in deep space, lower powers will probably prove most satisfying with these subjects.

Here are listed some of the most celebrated and readily seen examples; many belong to regions near the center of the galaxy, and thus are summer objects in the north and winter objects in the south. You can choose other examples, of course, but do not expect to find them evenly distributed throughout the sky. The main purpose here is to observe the various types of nebula and note the characteristics of each type.

Great Nebula in Orion M42 – This object is magnificent beyond description. Brilliantly lit, it is populated with countless tiny hot youthful stars within and around the Trapezium. One of the best-known stellar nurseries, huge clouds and swirls of illuminated emission gases, bright reflection components, and significant unlit dark regions cross and intersect throughout.

Omega Nebula M17 – This object appears as a large "swan" swimming in outer space! This emission nebula is one of the most striking and is mixed with dark unlit gas and matter as well as many young hot stars.

Trifid Nebula M20 – The famous triple nebula, of emission, reflection, and dark types, this is dissected by the dark components in lanes toward the center, where a six-component multiple star illuminates the reflection portions. Depending on various factors, you may or may not be able to resolve all six stars.

Lagoon Nebula M8 – Without doubt one of the prettiest sights in the sky, the Lagoon Nebula has great swaths of dark nebulous clouds cutting through it like a great lagoon or river. Adding to the magnificence is a stunning young cluster of newborn stars on one side of the "lagoon." Found here are a mixture of emission, reflection, and dark nebulae.

Eagle Nebula M16 – This object has regions of dark gases that create the exciting visual effect of folds within the fabric of the whole!

B143 in Aquila – This is a sudden, eerie dark hole in the surrounding star field.

B72 in Ophiuchus – This is the well-known serpent-like S-shaped dark nebula void.

B86, the "Ink Blot" – Here is another easy to see dark void at the edge of cluster NGC 6520 in Sagittarius. It is possibly the most startling dark nebula of all.

The Milky Way is studded with nebulae of all shapes, sizes, and types. Of the most brilliant telescopic objects, among our most magically mystical sights, what we see as separate nebulae are often just illuminated sections of a much broader whole. Many of the most celebrated nebulae are interconnected by vast dark portions, and what we cannot see constitute even greater nebulae! The largest nebulae are not likely to be brilliant or exhibit dramatic qualities, but they may justify the effort to see them.

Quick Project: Viewing Large Diffuse Nebulae

Time Required: 3–5 min per Nebula

With moderate and larger apertures, low-power large binoculars, or "richest field scopes," in decent conditions some grander scale nebula structures may be effectively observed, such as the following:

North America Nebula NGC 7000 – The resemblance to the continent is not as obvious as it seems it should be!

California Nebula NGC 1499 – This is very hard to see, but with patience you will, even from poor locations.

Rosette Nebula NGC 2237 – This also surrounds the glittering open cluster NGC 2440 and is a wonderful hollow cloud, appearing somewhat like an open flower.

In our own galaxy you will find an amazing array of very large diffuse dark nebulae as well.

Pipe Nebula B78 – In Sagittarius, the central hub of the galaxy, are some of the most stunning views of the Milky Way we have, including this one. Here are many dazzling fields of stars and dark nebulae, interacting in remarkable ways.

Be warned, however. The illuminated examples are large structures with low surface brightness; it is only their total luminosity that accounts for the magnitudes assigned to them. In the case of dark nebulae, you will need reasonably dark skies to see them convincingly. To see any of them really strikingly, you will need optimal conditions as well as really dark skies. As such, they are challenges for live viewing, but do not be deterred; they can be seen from suburban surroundings at that. Use every means in your arsenal, but do not expect them to be among the best sights from your home site, especially if you do not have the luxury of leisurely observing sessions.

There is also the chance that you will detect a trace of color within reflection nebula, as long as you have enough aperture and subject brightness. Because we do not have many opportunities to see coloration in the depths of space, this is truly an exciting prospect. Naturally, the choices are limited. You will need natural viewing, but some light filters help. If you have a moment, you might check out the following examples in this easily carried out project.

Quick Project: Seeing Colors in Deep Space

Time Required: 5 min per Nebula

Try to maintain a special alertness for color while you are observing the brighter nebulae – conventional or carefully filtered viewing only, of course! From optimal locations, not only you will see much more detail than otherwise would be the case, but there is the real possibility of making out more noticeable traces of different coloration. However, for the most part you will be limited to the brightest and most celebrated examples to see any suggestion of color at all. Some we have already viewed; now is the chance to very carefully look again at them for this other visual characteristic.

Trifid Nebula M20 – a glorious blend of both emission and reflection gases, is a good place to start; it is possible to see traces of pink and blue color in this nebula (within the red emission, and blue reflection sections), even from suburban sites!

Omega Nebula M17 – its golden-yellowish hue is striking, and easy to see.

Great Nebula – in Orion M40 may show you shreds of red, green, and bluish colors at the fringes.

Crab Nebula M1 – You may sense traces of reddish color in the faint tendrils from the suburbs with conventional viewing along with a narrowband filter! A remnant of a Milky Way supernova almost 1,000 years ago, it remains an elusive object from such locations at the best of times, so you will need good conditions and viewing skills.

When compared with the brilliant hues of our own world, these will seem faint indeed, and more likely to show only at the outermost fringes than deep within the structures; the degree of color we are talking about is nothing like you see in a typical CCD image! However, once again, the brain is capable of remarkable degrees of adjustment, and the effect of coloration may increase as you begin to notice it. However, do not anticipate easy results or frequent success with everything you try.

Planetary Nebulae

Emission nebulae include non-diffuse planetary nebulae, so-called (by Herschel) because of their frequent similarities in the field of view to the appearance of planets, both in size and shape. Their existence is, however, dependent on entirely different factors than star creation; often an almost undetectable central star provides the energy to light the nebula, exciting the hydrogen through ionization. Far from being stellar birthplaces, the opposite is true; in fact, this is the way many old stars spend their declining years. Being formed from single stars, most planetaries are fairly small in cosmic terms, although their brilliance as emission objects more than makes up for it.

There are three distinct ways to enjoy viewing these objects, each adding to your completeness of understanding, and there is an abundance of fine examples of planetary nebula throughout the sky in our local celestial neighborhood. Most can be readily seen in conditions of considerable light pollution and in short sessions at the eyepiece at that; for the busy amateur, they must surely rate very highly in viewing priorities. Almost all of them respond very favorably to electronically enhanced viewing, in fact, disproportionately well, compared with many other objects seen under poor conditions.

Quick Project: Viewing the Brightest Planetary Nebulae

Time Required: 5–10 min per Subject

Simple, unaided conventional viewing will always show something worth seeing. Planetaries will often look least spectacular in this manner, but are still likely to be impressive. A more striking view of many planetaries may be gained with the use a narrowband filter, which will suddenly throw them into stark relief, their emission gases usually showing as a luminescent blue. They frequently look their best this way, depending, of course, on the characteristics of the specific nebula itself. Planetaries respond in a most dramatic way to image intensified viewing and CCD video. These devices exaggerate the emission gases brilliantly, often giving the best structural views of the nebulae themselves. Nevertheless there are some notable exceptions. Numerous fine examples are available; here are just a few:

"Eskimo Nebula" NGC 2392 – This object seems to show the best of its glowing "face" with the most straightforward of viewing methods.

Dumbbell Nebula M27 – Responds beautifully to any form of viewing. In the conventional view, its ghostly white form is easy to see.

Ring Nebula M57 – Perhaps the most famous of all, this nebula easily reveals its "puff of smoke" form, the hallmark of its appearance. Great observing skills are needed to sight its subtleties, such as interior shading and banding, and especially the infamous central star.

Dumbbell Nebula M27 – By CCD video or image intensifier, this shows itself quite differently now (the great apple core in the sky!), not only in shape, structure, but in revealing crossed lines of superimposed stars, and especially the central illuminating star.

"Cat's Eye Nebula" NGC 6543 – Its helical winding is never more clearly seen than through a Collins I3; it seems that it needs the substantial extra boost in luminescence in order for the eye to readily discern this subtlety.

"Eye" Nebula – Or as it is often known, "the Ghost of Jupiter" NGC 3242, is quite wonderful under intensification; never does one feel stranger than when peering at the almost-real staring "eye" in space looking back, so clearly defined.

NGC 40 – Shows its surrounding ring and extensions in a way one will never see in the conventional view, when viewed with image intenfication.

The list goes on. Interestingly, often hard to detect central stars in these nebulae usually become obvious in enhanced viewing; there is hardly one that will not allow us an easy sighting. Without doubt, planetaries are among the most fascinating, varied, and plentiful of all deep space objects for amateur observers.

Ever-Deeper Space

By the time we exit our Milky Way neighborhood, we discover how the universe looks "on the outside." Here are glorious views for the astronomical enthusiast who is also fortunate enough to have equipment sufficiently large to access it (see Chap. 2). Beyond all of the visible stars punctuating the vault above is just an endless array of faint and fainter hazy patches in all directions. Each of these seemingly innocuous formations are complete and separate island universes in their own right, spread throughout the cosmos, similar gas and star formations such as we see in our local galaxies. Hundreds of billions of them! However, as observers, for the most part we must content ourselves with a limited number of relatively nearby galaxies. Those lying much beyond 100 million light years appear simply too small and faint to provide worthwhile viewing or study, especially when our time for observation is also limited. Luckily, there is no shortage of relatively nearby galaxies to see!

Although there is much pleasure to be had from exploring galaxies at the very threshold of visibility, and particularly when many lie near each other in great clumps, such as the wonderful Fornax or Virgo (Galaxy) clusters, there is only so much detail that can be extracted from viewing these galaxies, or ever more remote ones, since seeing detail within any of them remains unlikely. However, the best among the closer destinations invite countless hours of individual study, and may be successfully viewed with ordinary means. Certainly, our hopes of seeing spiral structure are maximized by relative proximity. Try viewing the Whirlpool Galaxy M51 if you want to be left speechless! The greater cosmos will easily occupy whatever time you have at the eyepiece, to the degree that it is inconceivable that much of this simple, practical, and awe-inspiring activity has been by-passed by many enthusiasts for something so impersonal as CCD imaging and the like.

Galaxies

Perhaps the single most enthralling quest in deep space is seeing galaxies revealed as spiral formations. Such visual delights were not in the realm even of the professional astronomer in the nineteenth century. Only in relatively recent times have amateurs had larger apertures of modern optical quality and other advanced equipment available to them, and thus had the opportunity to actually see such things at the eyepiece. This still seems utterly extraordinary. At decent observing sites, so many galaxies come within our grasp that it seems as if we will spend the rest of our lifetimes exploring *all* of them. This might be true except most of us, as busy people with full lives to lead, will probably not have the chance to visit these favorable observing places too often. Therefore, we should maximize any potential we have to view galaxies wherever we live, although seeing spiral structure itself is naturally harder at these typically less than ideal locations. From suburban/urban backyards, and even if your time is limited, you can still see many galaxies, but should realize that most of them will hang onto their spiral secrets until you can access better viewing sites. The exceptions are among the closer ones, most notably M33 and NGC253. You will find quite a few, in fact, as your perceptions evolve, that do, indeed reveal their makeup, albeit very faintly. This should be enough to keep you going! Just temper your expectations appropriately. You can also expect to see many interestingly detailed galaxies from your backyard, with edge-on galaxies (together with their dust belts), irregulars, and ellipticals often showing very effectively.

Using narrowband filters with most galaxies is not likely to be advantageous in dark skies. Galaxies usually do not respond as do emission nebulae! However, the same cannot be said of urban locations, where every aid against light pollution is valuable. Again, try viewing in the early hours when city lights are much reduced! Enhancing devices, such as image intensifiers and CCD video cameras, reveal many galaxies very well, even when light pollution is fairly high. Meanwhile, just remember that in all galaxy observing you must allow your eye to settle and adapt to the faintness of deep space; the forms will slowly begin to leap out at you. Few examples will be glaringly obvious at first blush. Above all, do not be daunted by limited opportunities to get to ideal locations; you will be surprised what you can see with just a little care and determination right from home.

Although galaxies respond dramatically to image intensifiers or CCD video, the frequency response of these electronic aids is typically skewed to the infrared portion of the spectrum. This is frequently quite useful when viewing galaxies edge-on with respect to us, whose dusty mantels often radiate powerful infrared wavelengths, and also to those of otherwise similar infrared spectrums, such as elliptical and irregular galaxies. However, there are no absolutes here; you may be astounded just as often as disappointed with face-on galaxies, which, theoretically at least, are supposed to be least responsive to such viewing. Frequently, these turn out to be spectacular subjects when conditions are dark and transparent. There seems to be no way to predict quite how any example will respond to enhancing devices, only to try and find out for yourself. You need to experiment, using your best educated judgment as your guide, and that is the best you can do. While you will find what you anticipated to be frequently correct, do not make the mistake of thinking that you can rely solely on your best instincts. You will often be wrong!

Quick Project: Viewing Detail in Galaxies

Time Required: 10–20 min per Galaxy

The brighter galaxies make ideal subjects for thrilling viewing, even when our circumstances and equipment may not be all we might wish. Even from highly light polluted suburban sites, many may be seen to satisfaction, although obviously under dark skies they come into their own. Here are just a few of the best examples of easily viewed galaxies of various types, all of which respond well to *any type of viewing method*, in reasonably good conditions. Surprisingly, many of the face-on galaxies listed here also rate among the best subjects for image intensified viewing:

Whirlpool Galaxy M51 – This face-on galaxy, relative to us, is a magnificent double spiral, with one arm appearing to reach out and join a smaller galaxy nearby, NGC 5195. When skies are dark enough we can see these features fairly readily with apertures starting at around 10 in. only (25 cm), thanks to modern optics; just knowing the galaxy's visible structure does not hurt either, but wait for the darkest opportunity.

Sombrero Galaxy M104 – The great "hat" in the sky makes for spectacular viewing by any means, but it never looks more like its photographs than when seen in the enhanced view (integrated CCD video or image intensifier). There is so much brilliance, clarity, and detail you will hardly believe what you are seeing. One of the grandest galaxies of them all, near edge-on, with a prominent dust lane, its dish-like form appears almost concave, and you will swear that you can trace the dust lane all around.

M82 – This irregular "explosive" galaxy is one of the finest sights in the sky and is loaded with detail. It is so bright, and so full of mottling and irregular structural features that it is hard to take your eyes away.

NGC 4565 – This is the most celebrated of all completely edge-on galaxies with a bright core and an extremely striking dust belt extending all around. Of magnificent dimensions in the field of view, this is a sight you will never tire of seeing.

NGC 253 – Near edge-on, this is a fine, grand spiral with very striking spiral form readily visible, even from poor locations. With much complex detail on parade, it will easily fill your field of view. Despite the fact that it lies at a low altitude in the Northern Hemisphere, it always puts on a magnificent show.

NGC 5128/Centaurus A – A phenomenal sight, this vast elliptical galaxy is colliding with smaller edge-on spiral and is one of the most impressive and complex galactic sights of all. Unfortunately it and nearby Omega may prove problematic, if not completely out of range, for many Northern Hemisphere observers, because it lies so far to the south. In this hemisphere, you will need to pick your observing times very carefully, regardless.

NGC 2903 – This is a striking example of a barred spiral, seen almost face-on.

M84 – This elliptical galaxy is bright in any field of view, but this type of galaxy also responds quite brilliantly to electronically enhanced viewing. It is unfortunate that this type is usually quite disappointing if you are hoping to see any structural detail, since their appearance is usually limited merely to seeing a bright blob by any method of viewing! Perhaps you will be able to get a glimpse of the famous "jet" shooting out from M84, which is within the realm of possibility. A problem in viewing this feature is that it requires some fairly high magnification; this may eliminate the most effective use of image intensifiers.

M106 – Another almost face-on galaxy. Although not having such strong spiral attributes as does M51, its twisted form nevertheless may be seen handsomely in moderate apertures.

NGC 891 – This is one of the finest fully edge-on systems; with a full width dust belt, it is nevertheless somewhat fainter than NGC 4565, but hardly second to it in overall splendor and not hard to see in fair conditions.

Pinwheel Galaxy, M33 – This magnificent face-on spiral is so close (next closest after M31) that it will fill the field of view even at low powers. Viewing the galaxy may require averted vision at first. Eventually its form will become evident, its two great arms reaching far across the field of view. You should also note the blotches of iridescent nebulae in them, which are conspicuous. This galaxy is one of the largest visually, and the second closest major galaxy after M31. From the best locations you can see it as a widely luminous whole.

This short list is sufficient to indicate just a little of what may be in store. Because these are particularly bright, outstanding examples, they are some of the best places to start. Not all galaxies you observe will be so ready to give up their secrets, however, and you will probably need patience to make out ever-greater detail. The extent of just what you see depends on the many factors we have covered earlier in this book, but a good equipment guideline would be always to choose your telescope around its potential to show you galaxies. Once again, this means aperture is largely the name of the game. (Do not forget that this has to be a *solidly mounted* aperture at that, or it will not realize its potential.) If a telescope shows galaxies well, then it is not too small for anything. Couple this to buying only the most appropriate and useful features on your telescope so that you do not waste your money on add-ons that do little for your viewing! At each level, buy well once, and not poorly many times over.

Novae, Supernovae, and Variable Stars

Sometimes a large star in our galaxy becomes unstable from various causes and develops so much energy output that its own gravity can no longer hold the star together. The result is a true cataclysm – a supernova – an explosion sometimes generating enough light to turn night into day throughout the entire galaxy. Less spectacular, far less dazzlingly brilliant, though still devastating by any standards, are novae, stars that experience vast explosions, but because they are not disrupted from their inner core go on to live another day, sometimes as changed objects, often only to explode again.

Because novae are "gentler" forms of stellar explosion, they are much more likely to be seen only within our own galaxy and can make spectacular events during the rare occasions in which they occur. Both of these phenomena represent a complete field of interest and study to those who passionately follow them. It is simply amazing that from very great distances across intergalactic space we are actually able to discern the light coming from just one star as it erupts into probable oblivion. Usually there are numerous examples to be seen in the visible universe at any one time, even if many, especially those in more distant galaxies, require CCD imaging or similar to capture. In any event, an aperture much less than 12 in. (30 cm) will make viewing most of them in real time a struggle, if not impossible.

However, enhanced viewing devices make many otherwise unseen or faint examples readily visible at the telescope. There is always the chance that you will stumble across something, especially when appearing in views with which you are very familiar. Live viewing of such mighty cosmic events surely represents potentially some of the most dramatic viewing we have.

The Crab Nebula M1, previously mentioned, is what was left from a supernova right in our own galactic neighborhood, at a relatively "close" 6,300 light years. Exploding in 1054, its brilliance must have been astounding from Earth, even visible in the daytime for almost a month, and awesome it still is, almost 1,000 years later. Because of its relative proximity, astronomers have a very good reference as to the nature of similar objects throughout the cosmos. At dark sky sites, the Crab's glowing tendrils (remnants of the star's outer layers) still speak of the power once unleashed by this stupendous blast. It remains a marvelous object in the eyepiece by almost any viewing method, even revealing quite easily with an image intensifier its postexplosion 16th magnitude neutron star remnant, as well as the shock wave of repulsed gas. Frame integrating CCD video cameras will show the star quite readily as well. Within the nebula are all kinds of detail. If you are able to see it live for yourself, remember that this tiny ultramassive star would fit into just a large lake on Earth, and yet, we are able to see it across such vast reaches of space. There have been astronomers who have spent much of their entire careers studying the Crab Nebula alone!

Although it is possible to locate all kinds of supernovae amateur web sites and scientific group sites, perhaps the most significant society concerned with all things supernova is the International Supernovae Network, at www.supernovae.net. You can certainly monitor the site regularly for readily visible examples exploding before your very eyes. Even a casual visit to it will generate some interest among the least likely to be inspired, with multitudes of images of recent and past events, and also a reference source to most nova and supernova websites and those related to them. However, be aware that a significant degree of involvement in this type of phenomenon will entail more time than you probably are prepared to give.

Perhaps you may also be interested in the observations of variable stars; if so, you might visit the website of AAVSO (the American Association of Variable Star Observers) at http://www.aavso.org for very specific information on all aspects of this specialized field of study. The variations in variable stars' brightness come about from two distinctly different physical causes: the eclipsing effect of one or more stars in mutual orbit around each other, and genuine variations in output of a star itself. However, again, this type of astronomy is unlikely to fit the lifestyle of anyone who is pressed for time, because it depends on fastidious and extremely careful observing and record keeping. There is nothing to stop you from observing a few of the best-known variables, such as Algol, which, over several sessions from time to time, should not prove too problematic, maybe even be a welcome change of pace.

CHAPTER FIFTEEN

Deep Space Imaging

When presenting anything astronomical in book form, illustrations of some of what is described naturally play a fairly central role. Despite the fact that we are not really astro-imagers, because of time limitations, a general guide to the results you might expect to see live in the eyepiece is an important ingredient. It is also likely that you will want to be able to record some images of your observations, perhaps to show to others but also to relive what you have seen. Deep space represents the ultimate challenge, and if time is not your friend, it is highly unlikely that you will be lured toward *any* of today's standard and commonly ordained methods. However, there is surely at least one of the alternative methods that you will be able to try for yourself. None of them requires a significant investment of time.

Eventually the shortcomings of every imaging technique used become increasingly apparent. Ordinarily, without resorting to full-blown CCD imaging, the more elaborate frame integrating CCD video applications, or even traditional astrophotography, it would not be feasible to go further than the most rudimentary methods. But the goal of finding the most effective, simple method for showing the way objects appear in the eyepiece remains.

Despite everything you may have read these days concerning the dominance of CCD imaging and its applications to deep space, it is perfectly feasible to keep your own imaging far simpler than this. It is quite acceptable to produce very effective and realistic imagery with only the simplest of means, such as drawing in lead pencil on a white background (a kind of "negative" imagery in the photographic sense that can later be reversed into a "positive" image). A more technical approach, with only a fraction of the effort required for CCD imaging, is CCD *video* imaging with multiple integrated frames. This method is capable of creating stunning pictures.

A. Cooke, *Make Time for the Stars: Fitting Astronomy into Your Busy Life,*
DOI: 10.1007/978-0-387-89341-9_15, © Springer Science + Business Media, LLC 2009

It also allows you to select the frames you wish to stack later on your computer, rather than utilize the integrating function of the cameras themselves. The potential is considerable. However, with each increasing technical application, the time commitment is naturally ever more demanding, not to mention more costly.

There is only one method that comes close to being truly instant, but it does require the use of a modern image intensifier. The device can be coupled either to a CCD video camera or digital camera, the latter being the method used to illustrate this volume. Unfortunately, as you are now all too aware, this is an expensive option, with the cost of the intensifier possibly exceeding that of many telescopes! However, the process is so good, so simple, and so quick that it may just be the answer for you, too, if you can lay your hands on such a device. The camera is attached with a standard adapter (by Celestron) to the image intensifier eyepiece. The focus can be adjusted essentially to infinity (sometimes a little fine adjustment of this parameter is required), and with a little experimenting you should be able to find the most effective combination of aperture, shutter speed, and contrast sensitivity. That is it!

With very short exposures (ranging from less than a second to 3 s at most), you can record the most stunning deep space imagery. Admittedly, it is monochromatic, but the detail and resolution is as good as many a CCD image taken over a long period of tracking. The brief exposure length serves a double purpose by averaging the electronic "noise and scintillation" in the intensifier's image to produce a smoothly realized whole. There is no processing required of these images, only the removal of the image intensifier's green hue (changing to black-and-white), if required. All of this could not be easier, and for the enthusiast who has very limited time, what could possibly be better?

But let us start by outlining other viable methods; any of them may work for you, too, and they would not make undue demands on your time.

Drawing

Many observers' original approach to deep space imaging was to develop the now ancient art of sketching at the eyepiece. Using both conventional viewing (with and without a narrowband filter), as well as sometimes views via an image intensifier or CCD video camera, it is possible to image many deep space objects with remarkable realism. This method is pretty quick, too, so for many observers it will still prove remarkably suited to them with the time constraints they have. Once familiar with the process of putting subtle lead pencil shadings onto the page, it is possible to obtain some exceptional likenesses, albeit in a photographically "negative" sense. When the image is scanned and reversed to become white on black, the effect is often remarkable. However, the fact that it is a subjective process to a lesser or greater degree, not everyone will readily accept the results as authentic. In any event, good as it may be, it is hard to claim that it produces entirely accurate results, because of the faintness of the subjects themselves.

Quick Project: Drawing Deep Space Objects

Time Required: 10–20 min per Object

The method is very simple. Very light strokes of soft lead pencil on plain white paper (basic computer stock is best), blending with fingertips, and shaping with an eraser is most of the technique required. (White pencil on black paper is far less satisfactory, since the medium is not nearly so receptive to subtleties.) As with everything in space, representing less on the page is more akin to the reality, so do not use pressure on the page to underscore anything. If carefully carried out, you will arrive at something like Fig. 15.1.

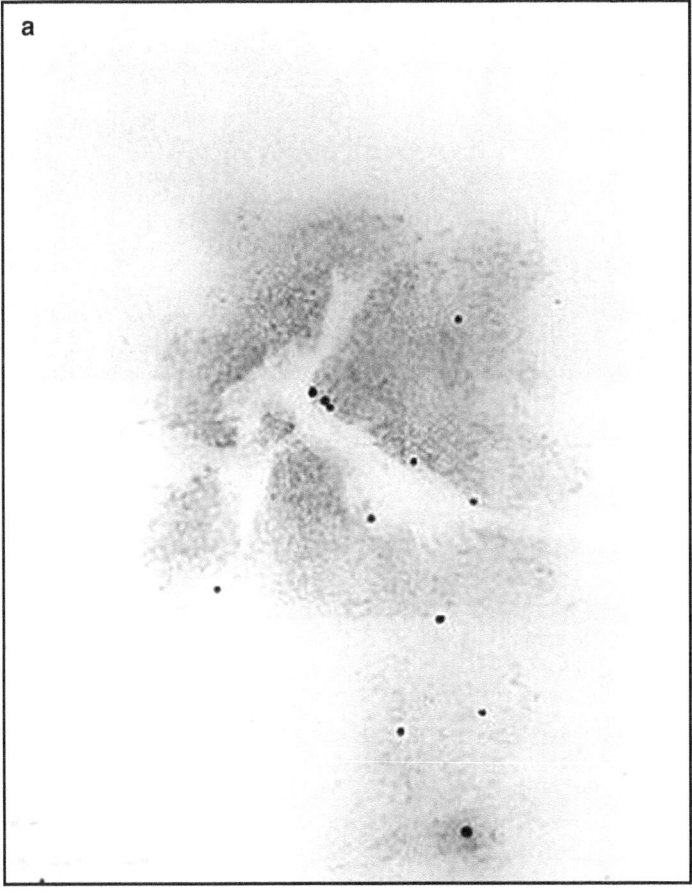

Fig. 15.1. (a) M20 – the Trifid Nebula (suburban view with a narrowband filter) and (b) •••.

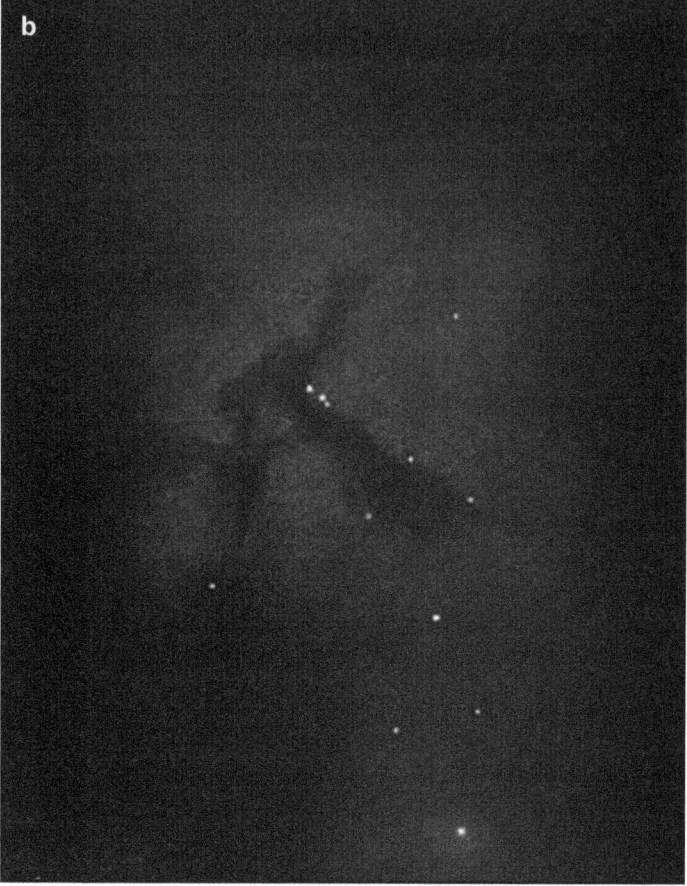

Fig. 15.1. (continued) The same image 'reversed'. (AC)

Simply take this image and scan it into your computer. Then, just reverse the image into "positive," a capability available in the most basic of programs. The final result may be judged by how effectively you feel it has represented what you saw in the eyepiece. It will immediately become clear when "more" is, in fact, "less."

Examine Fig. 17.9 in Chap. 17 to compare these illustrations with the results using a digital camera and an image intensifier to see how effectively this object is represented, individual stellar brightness differences notwithstanding (due to the response of the image intensifier).

Two more examples show even further what can be done by drawing, and how effective this simplest of all imaging methods actually is. It should be pointed out, however, that M82 (overleaf, Fig. 15.2) was taken in conjunction with an image intensifier, whereas the view of M20 was the result of conventional viewing with the addition of an Orion Ultrablock filter. Nevertheless, an image intensifier is not a prerequisite, but merely makes the galaxy far easier to see well from suburban home sites.

a

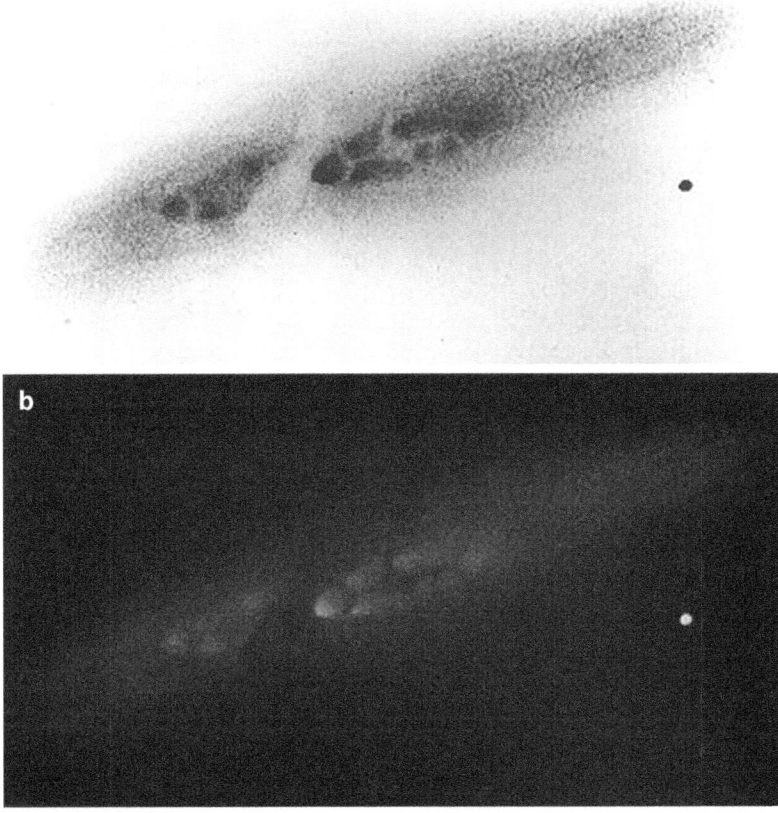

Fig. 15.2. (a, b) M 82 irregular galaxy.

Again, compare these pencil efforts with Fig. 17.6 in Chap. 17.

Certainly the ease of this process, plus the degree of immediate satisfaction from the imagery it produces is considerable, despite its limitations. If you are not interested in reaching greater realism, you can certainly stay with drawing indefinitely!

CCD Video Imaging with Image Intensifier

In an attempt to reach greater realism you might try extracting single frames from simple video streams (nonframe integrating), but it will require the use of an image intensifier (see Chap. 3). This system is fast, simple, and more effective than one would expect for creating instant deep space images. It also provides a pretty good method for group viewing around a monitor, one of the great advantages of CCD video applications. Nevertheless, even when using a recursive frame averager to smooth out electronic noise of the intensifier tube, it still misses quite a lot of the impact and presence of the live view. There is a certain quality that just does not translate to single still frames from moving video, despite the fact that the imagery is no longer subjective. No matter how carefully one goes about selecting or processing the individual frames selected, the effective resolution and luminescence such as the eye alone perceives just does not translate to them. Looking at many frames individually reveals that each contains different information; few are complete in every way. Nevertheless, it offers a new dimension, along with great ease in deep space imaging.

You may find yourself looking around again. Frame integrating CCD video is not a bad way to make further improved images, and it does not require the use of an image intensifier, either. Some imagery produced in this way is remarkable for sure and a wonderful solution for producing some pretty great pictures without spending a fortune. You will find many fine examples of such deep space images on the Internet, and as such they are remarkable, especially given the standards of what was possible in the past. With a tool as powerful as the new StellaCam II camera from Adirondack, some users have come remarkably close to the best amateur CCD monochrome imagery. Regardless, the built-in drawbacks (the upper limits of resolution) remain; we have to remember, after all, that it is only video. Additionally, while far easier than standard CCD imaging, it may be nevertheless more of a hassle compared with that of selecting single frames, or even drawing.

For those whose financial constraints place top notch CCD video or image intensifiers beyond reach, it is still quite feasible to use standard digital cameras (plus eyepiece adapter) in lengthy time exposures to achieve decent results. Of course, this will require accurate alignment of the telescope mounting, whether equatorial or computer controlled dual-axis tracking, which in itself can be time-consuming. However, this might nevertheless be the most straightforward approach for many such readers. Remember, too, that extreme accuracy of equatorial setup is hard to achieve, with the result that many enthusiasts do their best work while continuously monitoring the exposure throughout the length of exposure required; once again, it will require more time than many people may be able to find.

Bill Collins, of Collins Electro Optics (makers of image intensifier eyepieces for astronomy), has long been taking high-resolution digital camera deep space images via an image intensifier. You can see some of his images on the company website at www.ceoptics.com. According to Collins, these images consist of short time exposures (up to 6 s) using a basic Canon digital camera, taken through his 7-in. (17.5 cm) astrophysics refractor coupled to one of his image intensifiers. The fact is that a regular digital camera coupled to ever-greater apertures allows even better results!

The crispness and resolution of Collins' imagery comes much closer to the appearance of the live view than other quick systems, and the brief time exposures are sufficient to record amazing results with the aperture he uses. The combination of greater aperture, short focal ratio, and the substantial added power of the new Collins

Generation 4 image intensifier appeared to offer the best answer of all! The prospect, essentially, is being able to make high-quality snapshots of deep space objects. Plus, you will no longer need any type of recursive frame averager (as needed with image intensifier and video), because even short exposures have exactly the same effect! The system allows extremely short exposures on brighter subjects, yet only a little longer on less brilliant ones.

With the image intensifier installed in the telescope focuser and connected to the camera, you can set shutter speeds in appropriate durations and effectively time exposures. Depending on the subject brightness and angular size, the camera you select will need the specific types of adjustments previously listed. Zoom capability will be found necessary because the image field needs to fit the image frame, and small objects may need an even larger presence. Depending on the subject brightness and the degree of zoom function selected, these, in turn, limit your choices of focal ratio and shutter speed settings of the camera. This only further emphasizes the importance of manual control. A few minutes with a little experimentation will help immensely.

The resulting images require virtually no processing, so what you see on the page is basically how they will appear in the raw state, and essentially thus as in the live view. However, it must be noted that there are still additional qualities of luminescence at the eyepiece, and no amount of later touching up of your images will seem able to replace that special ingredient. Nevertheless, judge for yourself; it seems that in putting a higher quality image on the page, the idea has been extremely successful and certainly more representative of the live view than any other fast and simple method you are likely to devise. The best part is that it comes with none of the downsides of the other well-known imaging methods.

However, certain differences still remain between image intensified and natural eyepiece views. Because image intensifiers' spectral sensitivities (or even that of frame integrating CCD video cameras) are somewhat different to that of the eye, they will provide a corresponding visual emphasis toward their own specific spectral range. The differences are not so great as to render any object completely different, unfamiliar, or unrecognizable. Even in conventional viewing, different light wavelengths may be favored by special light pollution/transmission filters. Nobody would argue that the basic character of the object being viewed is significantly altered. The same applies to the intensified view, so using intensified images does indeed provide an excellent visual guide, regardless of your viewing method.

There are several important considerations in camera selection, among them being full control over the key functions of operation, and the small standard lens allowed for full use of the emerging light beam from the eyepiece. This negates the possibility of vignetting, something likely to occur with larger camera lenses. Specifically, features you will need, not necessarily available on many other similar cameras include manual control of focus, shutter speed, exposure time, time delay (essential to avoid magnified vibration of the telescope immediately after commencing the exposure), a large enough rear screen to attain fine focus fairly easily in the field, a wide range of exposure times, control to shut off the flash, as well as an ideal number of pixels for most uses.

It is an easy matter to attach and align the camera to the intensifier eyepiece with an appropriate adapter. Although many possibilities exist in the marketplace, you should consider the Universal Digital Camera Adapter by Celestron. This unit appears to be identical to other similarly named adapters, so you should be able to find one relatively easily. This well conceived, finely built, and inexpensive coupling device allows complete adjustment in all planes to align the camera precisely with the eyepiece lens. Best of all, it readily accommodates 2-in. eyepieces (hence, the 2-in. diameter Collins intensifier) without doing any damage to them because of its cushioned clamping. It could easily accommodate somewhat larger eyepiece sizes even than these!

Quick Project: Making Images with a Digital Camera

Time Required: 5 min of Setup per Object

Having first centered the object you wish to record, attach the camera with adapter to the eyepiece or an image intensifier viewer, and check or reset the camera's parameters. Be sure all components sit squarely relative to each other. Make sure the flash is off, especially if you are using an image intensifier! Take care to fine focus and center the image very carefully; it is easier when using an intensifier than conventional eyepiece. For the latter, focusing on a bright star or planet will have to suffice. It is easy in the excitement of the moment to overlook precise focus of the telescope on a small camera screen. Try a few exposures of different durations and settings for optimal results. With an image intensifier, it could not be quicker, and seems all *too* easy!

A Comparison of Methods

In conclusion, given here are three examples of images of the same object (NGC 253), made under similar circumstances and utilizing the various imaging techniques as just described. The dramatic improvement of each method over the prior one is immediately apparent (Fig. 15.3).

Visually, NGC 253 is large and impressive in the field of view, unusually bright for a galaxy, and features a wealth of lanes, mottling, and a striking, easily seen spiral structure. It is unfortunate that this magnificent sight lies so low on the horizon for observers in the Northern Hemisphere, which makes it less than ideally accessible. However, careful planning will reward you with one of the finest galactic spectacles of all. It is remarkable.

As in any form of imaging, you still need to understand what you are seeing on the page, versus through the telescope. In live viewing, especially with conventional eyepieces,

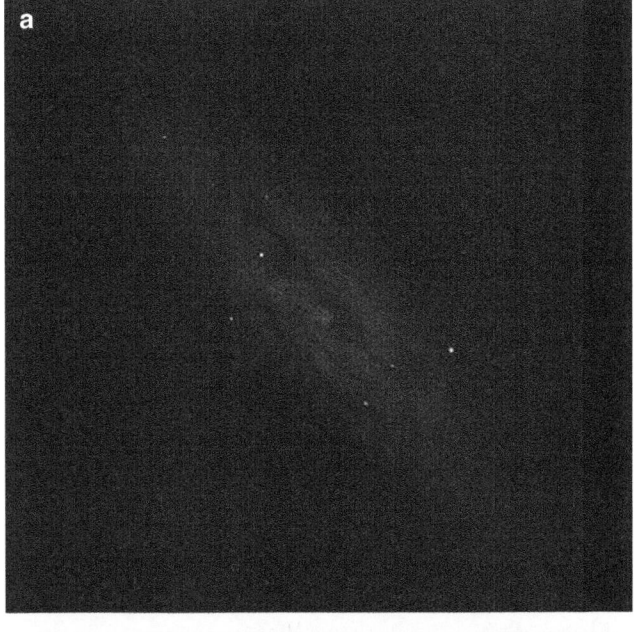

Fig. 15.3. (**a–c**) NGC 253: (**a**) Drawing with a narrowband filter, (**b**) Image intensifier and CCD video frame, and (**c**) Digital camera and image intensifier; 2-s exposure. (AC)

Fig. 15.3. (continued) (AC)

the eye and brain have an opportunity to compensate for extremely low light levels, which is all part of what occurs to a greater degree as one develops observers' skills. The actual true brightness becomes relative, and while the image may ultimately be very clear to us, we soon realize that what we are seeing could not possibly be photographed by any normal means. Sometimes, certain things show less prominently in the intensified view than "au naturale," sometimes more, but at least they can be easily photographed! Regardless, we need to make some subtle allowances at certain times. Refer to chapter 17 for a selection of deep space images produced by digital camera and image intensifier.

CHAPTER SIXTEEN

Astronomy via the Internet

There are probably many occasions when a little time on your hands does not necessarily translate into an astronomical viewing opportunity. Maybe the weather is not cooperating; perhaps it is daytime. More likely there is insufficient time to set up your equipment; just finding the energy to go through all that is entailed may seem like too much effort for just a few moments of looking skyward, no matter how great your enthusiasm. And sadly, not too many of us have the luxury of a permanently mounted telescope in some kind of observatory, ready to go at a moment's notice. However, such occasions offer something else instead: let us call it Internet Astronomy. With the current proliferation of Web sites on all aspects of astronomy, there is no shortage of exploration to do. With just a little dabbling, you will be able to discover a whole universe of information, presented as only this new worldwide resource can, as long as you avoid the obviously less than credible sources. It will only add value to your time at the telescope.

Gleaned from professional as well as many amateur observers, these sites feature the results of vast amounts of specific information and research concerning virtually everything in space you might want to see. They include catalogs of all types of objects, almost limitless perspectives gained from multiple sources, imagery from all levels of expertise, along with sites featuring all kinds of assistance for selecting and using equipment. All of this is presented in ways that few people would be able to track down in a library of astronomy books.

This is not to say that the Internet has replaced books! Far from it, since books are built around uniquely compiled and focused approaches. Rather, the Internet offers specialized forums for reference, if not necessarily presented with literary eloquence, or even a unified assemblage of ideas. However, here you can check virtually any statistic, observing time, apparition, magnitude, and so forth, ad infinitum.

A. Cooke, *Make Time for the Stars: Fitting Astronomy into Your Busy Life,*
DOI: 10.1007/978-0-387-89341-9_16, © Springer Science+Business Media, LLC 2009

Many of these sites have been referred to throughout this book. Nevertheless, putting it all together you might come to see Internet Astronomy as another form of exploration for you to enjoy.

In order to keep as relevant an approach to our needs as possible, specific areas of interest are categorized following, with brief descriptions in order to help make your searches as efficient as possible. Since so many of the sites listed link to so many others, it would be redundant to reference every site in multiple places. Thus, unless a related link or subsite warrants special attention (and there are quite a few that do!), a primary site is listed only once.

It is likely that you will quickly move beyond the guidelines here. It is really quite impossible to do this resource the justice it deserves in an overview such as this! Without so much as a telescope, you can visit other worlds in another way, and benefit from the sites' continual updates reflecting the latest imagery and information, and in ways that would have been unthinkable only a few years ago. Visiting these sites provides the ultimate "quick project," allowing you access and freedom otherwise unknown, and to suit any timetable.

The Moon

http://www.lunarrepublic.com

Interestingly, this Web site, operated by the Lunar Republic Society, declares on its home page that its mission is to provide an alternative to Hatfield's as well as Antonin Rukl's lunar atlases! It certainly does, so maybe a printed, fully realized version will eventually follow. The site features a grand photographic atlas of the entire visible surface of the Moon, plus a detailed catalog of every category of lunar feature. You will also find up-to-date information on lunar phases, links to other related sites dedicated to such topics as potential building technologies for lunar colonization, over 170,000 images of the entire surface from NASA's *Clementine* orbiter, and analysis of the lunar rock samples returned to Earth from the *Apollo* missions. Here too, the Moon is well represented with a logical and detailed order of presentation. By dragging your computer mouse over the many images, names of features are revealed.

http://www.apolloarchive.com/ apollo_archive.html

This site presents an overview of the entire *Apollo* program, with a wide range of information and comprehensive selection of photographs from the missions.

http://history.nasa.gov/alsj/frame.html

This link from the site listed next is dedicated exclusively to the complete archives of all NASA *Apollo* missions. Featuring all of the photographs taken, as well as extensive and full documentation of each mission, the site is one of the most extraordinary resources you are ever likely to see. The only problem is in finding the time to sift through the vast quantities of information! Certainly, no enthusiast, busy or otherwise, will ever run out of information here!

http://nssdc.gsfc.nasa.gov

This is NASA's primary archive, featuring a seemingly infinite range of information and topics. By typing whatever it is you wish to review into the top right hand corner box of the home page, you will be taken to Google with a huge selection of Web sites and detailed related subjects. These various pages again lead you to almost infinite Web locations, including many of the sites listed in this chapter.

http://nssdc.gsfc.nasa.gov/planetary/lunar/apollo.html

A link from its parent page, http://www.nasa.gov, this is another definitive source of information and images. Although featuring virtually links to the complete archive of all space research and imagery acquired through the space program, you will find access to some of it less straightforward than that found on the previously mentioned NASA historical site at http://history.nasa.gov/alsj/frame.html.

http://nssdc.gsfc.nasa.gov/planetary/online_books.html#moon

This is quite a remarkable find, and another division of the previously listed site. Apart from the many books referenced on the site, there is also a section devoted to on-line books, and they are free to download. You might be surprised how wide and generous the selection is - courtesy of NASA.

http://spaceflight.nasa.gov/gallery/images/apollo

This NASA site, dedicated to human spaceflight, will link you directly to http://images.jsc.nasa.gov, where you will find a comprehensive range of selected imagery from all past missions, including those of *Mercury, Gemini, Apollo,* and the space shuttle. However, for astronomical purposes, you may find the site less helpful than those with the complete archives already listed.

http://lunar.arc.nasa.gov/science/atlas/menu.html

This site features detailed mapping with the many different categories of surface feature, each accorded a separate "button" and all a direct result of the lunar exploration programs of all types. The site was still under construction at the time of this writing but looks promising as a reference at the telescope.

http://aa.usno.navy.mil/data/

A resource for observing, this site provides precise information on the Moon's placement in the sky and phase (as well as the planets), for any location, any date and time, worldwide.

www.astroplanet.info

This is a concise and accessible site with monthly information on the Moon and planets as well as charts for their positions and an archive for recent years.

The Sun and the Planets

http://www.spaceweather.com

This is the site specifically for sun/sunspots news, and current events in the Sun-Earth environment. For the dedicated solar observer, this may be the definitive place to go.

http://www.alpo-astronomy.org

This is the site of the Association of Lunar and Planetary Observers, one of the oldest and most venerated organizations for the amateur observer in the USA. Here you will find sections for the Sun, Moon, and each planet, with a vast compendium of images taken by its members.

http://www.nineplanets.org

This extraordinary site (in many ways the best pure observer's solar system site) contains as extensive a survey of materials as you could ever wish. Detailed backgrounds on every aspect of each planet is provided, including all relevant observational and scientific histories, along with a treasure trove of links to other related sites and imagery.

http://www.wwu.edu/depts/skywise/planets.html

Another very good site, with a slightly different approach, including a useful table showing all the planets' positions for any location at once, along with rising, setting, and transit times.

http://www.solarviews.com

In many ways similar to http://www.nineplanets.org, this site sets out to provide a wide-ranging set of tools to answer virtually any question about the solar system. Gleaned from every available source, there is more than enough material here to occupy many idle moments or hours of downtime.

http://jpl/nasa.gov

Great space exploration site, with links to all of NASA's missions, past and present; this is a huge resource on astronomy realized through spaceflight technology.

http://masil-astro-imaging.com/Don%20 Parker.html

Donald Parker seems equally at home in all forms of astronomical imaging, but it is his planetary portraits that stand head over shoulder above the efforts of others. Although no method seems still quite able to capture that special "living presence," he certainly seems to come closer than anyone else. On this site you will find a comprehensive selection of images, although solar system subjects (not including the Sun) dominate the site!

http://www.astrosurf.com/cidadao/index.htm

Portuguese Antonio Cidadao has made quite a name for himself in the international community of Moon and planet gazers, and not without good reason. The circumstances from which he observes and records his images would make many a purist shudder, but the results speak for themselves. Although perhaps not in Don Parker's league, Cidadao has certainly claimed his own piece of "astro turf."

Www.marsrovers.nasa.gov/home/

This is the site for the famed Martian rovers, Spirit and Opportunity. Quite apart from being able to view every image taken over the course of years, the site features information and insights on their missions.

http://messenger.jhuapl.edu

Here is the official NASA resource on the *Messenger* Mission to Mercury, along with all the latest detailed imagery.

Comets

http://www.space.com/comets

An excellent division of the larger parent site (http://www.space.com), this location provides much information, both general and specific. It features many images, including those taken during space missions in close proximity to their subjects.

http://www.windows.ucar.edu/tour/link=/comets/comets.html

Featuring multilevels from beginner to advanced reader, the site contains many links to images, with historic and scientific data.

http://stardust.jpl.nasa.gov

The complete story of the mission that recovered samples of matter from a comet.

http://www.solarviews.com/eng/comet.htm

The division of the larger site, dedicated to comets.

http://www.nineplanets.org/comets.html

Another division of a larger site, this portion has some general information, but only details on a few comets.

The Milky Way Galaxy

http://www.seds.org/messier

This is a wonderful and fully comprehensive site on the Messier objects, perhaps the best on the Web. Full of highly specific information on all of the objects, together with exactly where to find them, magnitudes, animations, even Messier's original catalog and descriptions, tables, and many detailed references and subsites, the site is also crammed with as many links as possible. For the amateur, this is an indispensable place to spend some time, not to be missed.

http://zebu.uoregon.edu/messier.html

Although not containing the complete Messier objects, the site has a selection of the catalog, featuring images of varying quality and text. Well worth a visit.

Variable Stars

http://www.aavso.org

This is the Web site of the American Association of Variable Star Observers; should this be an area of astronomy that is of interest to you, this is possibly the best site to visit.

Deep Space

http://hubble.nasa.gov

This is the official site of NASA, and although you will find a multitude of images here, they are by no means exhaustive.

http://hubblesite.org

Maintained by the Space Telescope Science Institute at the Maryland Science Center, this is more accessible and colorful than the previously listed site, with a wider range of images to study than NASA's, and organized into specific categories of object. Of course, Hubble's images extend to more categories than deep space, and you will find plenty to see regardless of your primary interest.

http://hubble.esa.int/science-e/www/area/index.cfm?fareaid=31

An exceptionally wide range of images from Hubble may be found here at the Web site of the European Space Agency, differently organized but virtually the equal of http://hubblesite.org, above.

http://clifty.com/scott/DSO

Here is an amateur site with deep space reference sketches, alongside the original black on white "negative" images. Although better drawings certainly exist, the site is a bold attempt to demonstrate the value of drawing, even in this day and age. There is a fair selection of objects (with some surprising omissions!), but a visit here should inspire you as to what may be undertaken with very limited means and the simplest of approaches.

http://www.members.aol.com/arpgalaxy/index.html

A wonderful compendium of irregular galaxies, with useful insights and detailed information.

http://antwrp.gsfc.nasa.gov/apod/astropix.html

In addition to checking out the image of the day, be sure to visit the archives. Here is a stupendous display of almost countless images of locations across the universe, from our celestial backyard to the truly remote. Not to be missed!

www.starrynights.com

The site of the ASD Planetarium – the site features interesting topics and links, though it is by no mean exhaustive.

Supernovae

www.supernovae.net

This is the granddaddy of all supernova sites, with vast information second to none. Images of all the latest phenomena are included, along with extensive departments and links.

http://www.sehgal.net/astro.htm

This is the site of a prolific amateur researcher, with a primary "bent" toward supernovae discovery. His impressive observatory would have rivaled many a professional institution just a few decades ago. Here you will find images he has recorded, as well as a comprehensive array of links to other sites on all aspects of astronomy.

http://cfa-www.harvard.edu/iau/lists/ RecentSupernovae.html

This listing of all supernova discoveries of the recent past (more than a year) is exhaustive and features complete information on each, including their exact coordinates.

Observing

Weather http//www.cleardarksky.com

This is the best site of its kind, which regretfully is not able to provide information on areas beyond the North American continent. Details about specific atmospheric conditions of concern to astronomers are the hallmark of this site.

Equipment

The number of sites on this topic is endless. However, here are a few sites that deal specifically with unusual equipment mentioned in this book.

www.ceoptics.com

The company Web site for image intensifier eyepieces, Collins Electro Optics is the sole manufacturer of these specialized devices.

http://www.cloudynights.com/ubbthreads/ showflat.php/Cat/0/Number/2053144/ page/1/view/collapsed/sb/5/o/all/fpart/1

This is an interesting page of a discussion forum, dealing here with the Collins image intensifier eyepiece.

http://www.weatherman.com

The creator of this page, Todd Gross, has provided a very informative site full of specific information on many types of astronomical equipment for the amateur, including telescopes, eyepieces, binocular viewers, filters, etc. He also includes pages on observing, astrophotography, his own images, and related astronomical topics such as the care of optical components, filters, and observing tips.

www.astrovid.com

This is the site of a manufacturer (Adirondack Video Astronomy) of CCD video cameras. Although Adirondack is primarily concerned with imaging, you will find quite a comprehensive range of astronomical equipment in general here.

www.jimsmobile.com

This is the site of JMI telescopes and equipment, makers of unusual and specialized instruments of the highest quality.

Filters
http://sciastro.net/porta/advice/filters.htm

This compendium of virtually every type of filter, along with discussions on their use, is not necessarily in line with the opinions put forth in this book, but may well be worth your time to explore.

Miscellaneous

http://www.nasa.gov

This is an interesting place to spend a little time. Much more oriented toward space missions and the knowledge obtained via this means of research, it also seems more general than specific. The site features press and policymaker releases and a wide range of information and imagery about every area of the universe. Although not everything is in line exactly with amateur astronomy per se, it is certainly closely related to that interest.

www.jpl.nasa.gov

This site contains the archives of NASA's many solar system missions, with a tremendous array of imagery collected over several decades. Again, it resembles in many ways the previous listed site, except that it is completely from the perspective of the Jet Propulsion Laboratory, where NASA's unmanned missions are developed and monitored. JPL has been intimately connected with the Space Age from the very beginning.

http://www.britastro.org/baa

Popular amateurs' resource, with images, drawings, sections for observing everything plus links to other resources.

http://www.space.com

An outstanding and multilevel site, this contains in-depth presentations of space science and astronomy in space.

http://www.worldwidetelescope.org

At the time of this writing, Microsoft has announced a new Web site, called Worldwide Telescope, which assembles a vast array of imagery from some of the most advanced spacecraft and observatories on virtually all subjects. It allows the user to access objects in space with a new flexibility not yet available anywhere else. Unfortunately, the software, free of charge, will be available only for Windows.

Robotic and Manned Spaceflight

http://spaceflight1.nasa.gov/gallery

This is NASA's human spaceflight site, part of http://spaceflight.nasa.gov, a site dedicated to providing a comprehensive overview and selected imagery of manned spaceflight from the beginning. You can pick any mission and enjoy a short history and selected photographs taken throughout the mission.

http://history.nasa.gov

The home site for records and links to all NASA missions, as well as the history of the organization and its personnel.

http://www.space.com

An excellent site for everything you might want to know about spaceflight, space sciences, and missions.

http://spaceflight.nasa.gov/realdata/ sightings/index.html

As part of NASA's Web site committed to human spaceflight (http://spaceflight. nasa.gov), here is a fascinating resource showing where and when to look in the night sky, for any location, to see such things as the International Space Station (ISS) or space shuttle. Although these spacecrafts may be seen without any telescopic aid, they may be difficult to track with most conventional mountings. However, there are some commercial telescopes today (of the dual-axis altazimuth variety) featuring dedicated software specifically for tracking the ISS. Maybe updates will become available for other spacecrafts in the future. If such man-made space objects are a major interest for you, then you might definitely consider such tracking capabilities essential for observing.

http://nssdc.gsfc.nasa.gov/planetary/ online_books.html

This is the complete listing of free offerings from NASA, covering the entire history of solar system exploration, including *Apollo*. Also included are historical writings, including the famous "Mars" by Percival Lowell, as well as numerous other related topics dating back to Newton.

CHAPTER SEVENTEEN

A Guide for Viewing Sessions

There is one ingredient that will greatly enhance those occasions when you are able to take part in your hobby: using such times in an organized and effective manner. This is much more important than it sounds; whenever you fail to organize your sessions, no matter how brief, you will pay a steep price, not only in the productivity of the session, but most notably in missing key objects that lay within easy reach. Regardless of how much time you have to spend, or your observing location, the need remains the same. Just taking a little trouble beforehand will pay large dividends.

Bad planning is even more noticeable when you finally can find time to get away. Because of all that is involved in relocating your equipment, it might entail a full day and night. Regardless, it is never easy to do, and always a hassle. If it is only going to be every so often, then at least you need to make it worth your while. Thus, every time you venture into the wilderness for a night of observing, you should try to have a clear idea of what you want to see, and also some inkling of what to expect. Such an approach will pay huge dividends in the quality of your observing sessions. It is also *much* more fun than aimlessly wandering among the stars with only a few destinations in mind, and "making things up" as you go. There is nothing worse than fumbling through your collection of field books in the dark (probably also in the cold or damp), seeking out objects to view, while trying to read from the pages with a red light. And then you discover that something wonderful that you could have viewed is now below the horizon!

A good way to proceed is to take with you a carefully organized catalog that you can use as a guide for most viewing sessions. Print it out ahead of time in a large, bold type that makes for easy reading in the red light that astronomers use for dark adaptation. Contained within this chapter is a listing of objects you can view effectively in the field. While fairly far-reaching, as an organizer and prompter it certainly is not intended to be a complete viewing guide. It is primarily concerned

A. Cooke, *Make Time for the Stars: Fitting Astronomy into Your Busy Life,*
DOI: 10.1007/978-0-387-89341-9_17, © Springer Science + Business Media, LLC 2009

with objects that are satisfying to observe, allowing easy imaging along the lines previously described.

Assembled over a long time period, the list provides just enough information, and sometimes just a little description, to trigger recall from prior observations. Intended for telescopes of moderate and larger apertures, the objects in this particular listing belong to all standard categories, according to placement in the sky rather than category. For general viewing, you may find more value to combining *all* object categories into the standard list, simply because there are many times during any night when certain types of objects are not in abundance; it is nice to have a lot of good options at any one time, especially if you are trying to maximize your observing sessions. If you are looking for fainter deep space objects, or a more specialized category, prepare a special list ahead, going about it in the same way, that is, in chronological sequence. Obviously, there are many other objects you could include (or some you might even exclude) in any list of your own, depending on your observing predispositions and equipment.

Another good step before you set out on your observing sessions is always to read up all that you can on those objects that you particularly want to see. The more you know what to expect and what constitutes their makeup, the more likely will be your success in viewing them, assuming you remain objective and do not "see" things that are not there! You will find enough here to keep you happily occupied for a long time, and only huge apertures will dramatically increase the possibilities, although, of course they will enhance what you see proportionately.

Interspersed throughout the listing are images taken utilizing an 18-in. telescope, Gen. 4 image intensifier and digital camera, with 1–4-s exposures, along with brief comments on the featured objects. They are included as a guide to what you might reasonably expect to see at the eyepiece through a moderate size telescope in dark sky conditions with conventional viewing refer to chapter 15. It is not necessary to use an image intensifier to see these objects in the manner shown. The majority of objects listed here will respond favorably to significantly smaller apertures as well. The difference at the eyepiece between ever-larger scopes is brilliance, increasing resolution and detail attainable, and better effective image scale. Sometimes, smaller optimum scales produced by lesser apertures will compensate quite well in an object's overall appearance, although creating *greater* image scale in these scopes soon negates this benefit. The amount of finer detail is thus proportionate to aperture.

You might start out by using the chart presented here as your own basic viewing plan for observing sessions, although ultimately you may compile your own. Depending on the time of year and day, you can begin referencing the list here at any point. Continue through your observing session in the order the objects are listed, since they are arranged *almost* in exact chronological sequence as they rise above the horizon (Herschel's numbers do not always precisely correspond to this). Objects are listed, regardless of latitude and hemisphere. Not everything will be visible from your location. However, although the intention was to make the chart as comprehensive as possible for all locations, including both the North and South Hemispheres, irrelevant objects may easily be deleted or passed over (check the declinations relative to your own). The chart was assembled from observations made from many locations, and so it is certainly relatively complete in its approach. You may also wish to arbitrarily skip many objects in the chart, only visiting those of one type, brightness, and so forth.

For the amateur observer, the imagery contained in many astronomical books is entirely unrealistic for visual expectations, very likely leading to disappointment. However, the pictures here are designed to serve as more of a valid guide as how the universe will actually look to you through your telescope. The object was to put you firmly in the observer's chair, but profound insights on what you see were not the purpose in this chapter. Aside from brief comments on each object showcased, any amount of information on these objects is readily available should you wish to research anything further. (The simplest way to go about this is online; just type any Messier or NGC number into a search engine, and you will immediately have access to a veritable universe of information.)

With the right circumstances, you should be able to see most of these sights for yourself with not too much difficulty. You do not need fancy hardware, a huge telescope, image intensifiers, or even specialized video cameras. However, you do need at least reasonable aperture, good optics, developed viewing skills, and dark, transparent viewing conditions for the best results of all, and certainly to see the fainter objects.

Each object in the following chart is listed with, where applicable or available, its catalog number first (NGC or other).

Object	Type	Magnitude	Angular size	Coordinates	Constellation	Comments
NGC 40	Planetary	10.5	60″ × 40″	00130n7232	Cepheus	Prominent ring with subtle ansae, central star, 11.5m. central star
NGC 55	Galaxy: Irr. or SBp	7.8	25″ × 4″	00149s3911	Sculptor	In some ways like M82; would probably be considered equal were it not so low in the sky during times it is visible
NGC 104	Globular "47 Tuca-nae"	4.5	25″	0024s7200	Tucana	Second only to "OMEGA CEN-TAURI"
NGC 205	Galaxy E6	10.8	8″ × 3″	00404n4141	Andromeda	M31 companion dwarf galaxy
NGC 221 (M32)	Galaxy E2	9.5	3.6″ × 3.1″	04270n4052	Andromeda	M31 companion dwarf galaxy
NGC 224 (M31)	Galaxy Sb	5	160 × 40	00427n4116	Andromeda	The nearest major galaxy; extremely large and bright but not resolvable visually into stars by any means

(continued)

Object	Type	Magnitude	Angular size	Coordinates	Constellation	Comments
NGC 246	Planetary	8.5	4″ × 3.5″	00470s1153	Cetus	Large and diffuse
NGC 247	Galaxy Sc	10.7	18″ × 5″	00471s2146	Cetus	
NGC 253	Galaxy Sc	7.0	22″ × 6″	00476s2517	Sculptor	Spectacular large, near edge-on galaxy; much detail and spiral form visible. See images in Chap. 14, Fig. 14.3
NGC 281	Emission nebula "Pac Man"	7.4	23″ × 27″	00528n5636	Sculptor	
NGC 288	Globular	7.2	10″	00528s2635	Sculptor	Relatively sparse
NGC 292 "Small Megellanic Cloud"	Galaxy Irr	1.5	3.5″	00530s7250	Tucana	
NGC 300	Galaxy Sc/Sd (S-shape)	11.3	21″ × 14″	00549s3741	Sculptor	
NGC 362	Globular	6	10″	01030s7050	Tucana	
NGC 404	Galaxy EO/SO	11.9	1.3″ × 1.3″	01094n3543	Andromeda	
NGC 488	Galaxy Sb	11.2	3.5″ × 3″	01218n0515	Pisces	Very compact; nearly face-on
NGC 598 (M33) "Pinwheel Galaxy"	Galaxy Sc	6.5	60″ × 40″	01339n3039	Triangulum	Large; hint of spiral structure even from poor locations, bright nebula NGC 604 visible

M33 is the next closest large spiral galaxy after M31 in Andromeda (Fig. 17.1). Fortunately, it is displayed face-on and we are able to view its full form in low to moderate powers. Loaded with bright knotty emission nebulae, such as NGC 604 (at the 11 AM position in the image below), make tracing the spiral shape easier. The galactic halo is quite evident, making this one of the grandest destinations we have. However, some observers have always reported failure to make out much at the eyepiece, so a carefully prepared approach is recommended beforehand, as well as the technique of indirect vision. Try it with this image. M33 demonstrates ably that galaxies are much very much fainter than we may have been conditioned to expect, even when seen from such close range. For those accustomed to CCD images only, take note!

Fig. 17.1. M33, the "Pinwheel" galaxy. (AC)

NGC 628 (M74)	Galaxy Sc (face-on)	11	9" × 9"	01367n1547	Pisces	
NGC 650 (M76)	Planetary "Little Dumbbell"	11	140" × 70"	01424n5134	Perseus	Irregular shape; reminiscent of M27; less so under scrutiny
NGC 869/ 884 "Sword Handle" Double Cluster	Two open clusters	7, each	35" each	02190n5709/ 02224n5707	Perseus	In near proximity; spectacular in moderate apertures
NGC 891	Galaxy; edge-on Sb	12.2	12" × 1"	02226n4221	Andromeda	With equatorial dust lane; wonderful object; fainter version of NGC 4565

Fig. 17.2. NGC 891. (AC)

Of the two grand dust-belted edge-on galaxies readily available to the amateur observer (NGC 891 and NGC 4565), in many ways NGC 891 (Fig. 17.2) is the more beautiful of the two, although significantly fainter. However, the setting of stars and its satisfyingly revealed form make it more magical in the field of view; at least it seems that way to me. Careful viewing reveals mottling along the dust

NGC 1023	Galaxy E7	11	8.6″ × 4.2″	02404n3904	Perseus	Lens shape with satellite galaxy on E. edge
NGC 1039 (M34)	Open cluster	6	20″	02420n4247	Perseus	
NGC 1068 (M77)	Galaxy Sb	10	2.5″ × 1.7″	02427s0001	Cetus	Bright, some spiral structure visible
NGC 1097	Galaxy SBb	10.6	9″ × 5.5″	02463s3016	Fornax	Some detail and structure visible
NGC 1232	Galaxy Sc	10.7	7″ × 6″	03098s2035	Eridanus	
IC 289	Planetary	12	45″ × 30″	03103n6119	Casseopeia	15m. central star
NGC 1291	Galaxy SB	10.2	5″ × 2″	03173s4108	Eridanus	
NGC 1300	Galaxy SB	11.3	6″ × 3.2″	03197s1925	Eridanus	
NGC 1316	Galaxy SO	10.1	3.5″ × 2.5″	03227s3712	Fornax	With tiny NGC 1317 12.2m. Galaxy SB 0.7″ ×0.6″

NGC 1333	Reflection Nebula	11	9″ × 5″	03293n3125	Perseus	Illuminated patch of otherwise vast dark nebulous region
NGC 1360	Planetary	9.4	6″ × 4″	03333s2551	Fornax	Very diffuse planetary; 9m. central star
M45 The "Plaiedes"	Open cluster with blue/white nebulosity	1.2	110″			Beautiful blue stars; very open; best with extremely low power
IC 351	Planetary	11	8″ × 6″	03475n3503	Perseus	With15m. central star
NGC 1398	Galaxy SBb	10.7	4.5″ × 3.8″	03368s2630	Fornax	
NGC 1399	Galaxy EO	10.9	1.4″ × 1.4″	03385s3527	Fornax	Brightest of 9 total: Fornax cluster
NGC 1499 "California Nebula"	Emission Nebula	4.5	145″ × 40″	04007n3637	Perseus	Extensive, primarily photographic, difficult to see with telescopes; use large binocular with low power scope
NGC 1501	Planetary	12	55″ × 48″	04070n6055	Camelop-ardalis	13.5m. central star
NGC 1514	Planetary	11	120″	04092n3047	Taurus	With 10m. central star
NGC 1535	Planetary	9	20″ × 17″	04142s1244	Eridanus	With 11.5m. central star
NGC 1553	Galaxy SO	10.2	3″ × 2″	0416s5540	Dorado	
NGC 1566	Galaxy Sb	10.5	5″ × 4″	0420s5450	Dorado	
NGC 1788	Reflection Nebula			05069s0321	Orion	
NGC 1792	Galaxy Sc	10.7	3″ × 1″	05052s3759	Columba	
IC 405	Reflection/ Emission Nebula		18″ × 30″	05162n3416	Auriga	Faint nebulosity, with var. star AE Aurigae
IC 410	Emission Nebula		20″	05226n3331	Auriga	Nebulosity surrounding cluster NGC 1893
NGC 1904 (M79)	Globular	8.4	7.4″	05245s2433	Lepus	Faint
Large Magel-lanic Cloud Dwarf gal-axy with no designation	Galaxy Irr.	1	6″	05200s6900	Dorado	Contains many objects: see Burnham
IC 418 "Spirograph Nebula"	Planetary	8	14″ × 11″	05275s1242	Lepus	Very bright (oval/bright 11m.) star
NGC 1912 (M38)	Open cluster	7.2	20″	05287n3550	Auriga	
NGC 1931	Reflec-tion and Emission Nebula			05314n3415	Auriga	Compact, "comet-like" nebulosity reminiscent of "Trapezium," with four illuminating stars
NGC 1952 (M1) "Crab Nebula"	Supernova remnant	9	5″ × 3″	05345n2201	Taurus	Resolution of tendrils possible with sufficient aperture

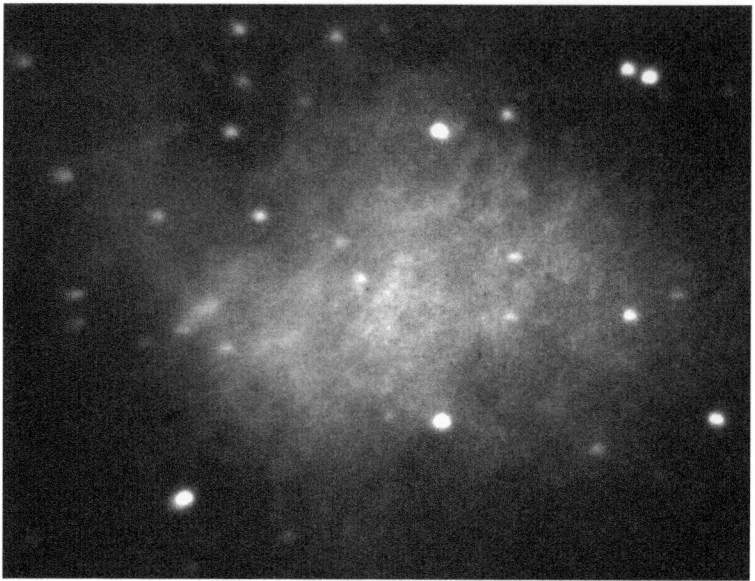

Fig. 17.3. M1, the Crab Nebula. (AC)

belt, indicative of emission nebulae. The view here, while made via image intensi-fication, is quite typical of the conventional view as well. It would seem that NGC 891 looks much as our own galaxy would appear from a similar vantage point in space.

Famous through the last millennium, the "Crab Nebula M1" is the remnant of the gigantic supernova of 1066, which lit up the night sky like day (Fig. 17.3). Look just left of center of the image and you will see the incredibly small, but unmistakable, neutron star at its heart, together with the well-known shock wave of receding gas - seen as a small bright arc at its right (at the "4 o'clock" position in the center). Not normally observed at the eyepiece, this should be an indication of just what the combination of substantial aperture and a Gen. 4 image intensi-fier can offer.

NGC 1960 (M36)	Open cluster	6.8	12″	05361n3408	Auriga	
NGC 1976 (M42) "Great Nebula in Orion"	Emission/ reflection nebula	5	65″	05354s0527	Orion	Exceptional; finest in N. Hemisphere; Huygenian Region new stars

NGC 1977 "Running Man"	Reflection nebula		40" × 45"	05351s0444	Orion	Bright; adjacent
NGC 1981	Open cluster	5.4		05352s0426	Orion	Near M 42
NGC 1982 (M43)	Emission/ reflection nebula		20" × 15"	05356n0516	Orion	Adjoins M42
S 2-240	Supernova remnant		2"× 3"	05360n2800	Taurus	
NGC 1999	Emission/ reflection nebula		16" × 12"	05365s0642	Orion	With 10m. star
NGC 2022	Planetary	12	25"	05421n0905	Orion	14m. central star
NGC 2024 "Flame Nebula"	Emission nebula		20'	05407s0227	Orion	Connected to "Horsehead Nebula" region
IC 434/B33	Emission nebula/ dark nebula		60" × 10"	05410s0224	Orion	Includes famous "Horsehead Nebula" B33; very difficult visually; try special "Horsehead Nebula" filter from Lumicon
NGC 2068 (M78)						
NGC 2070 "Tarantula Nebula"	Bright Reflection Nebula	8	8" × 6"	05467n0003	Orion	
NGC 2099 (M37)	Emission Nebula		20	0538s6900	Dorado	Extraordinary object within "Large Magenanic Cloud"; rivals even M42
IC 2149	Open cluster	6.4	20'	05524n3233	Auriga	Densely populated
NGC 2123	Planetary	10	10"	05563n4607	Auriga	14m. central star
NGC 2168 (M35)	Emission/ reflection nebula		10" × 10"	05416n0216	Orion	Fairly bright; part of Orion complex
IC 443	Open cluster	5.5	30"	06089n2420	Gemini	Adjacent to NGC 2158
NGC 2237/ 2244/2440 - "Rosette complex"	Supernova remnant		25" × 5"	06169n2247	Gemini	Curved arc; supernova remnant
NGC 2261 Hubble's "Variable Nebula"	Open cluster and nebula	5.5	80" × 60"	06323n0503	Monoceros	Impressive; center of the cluster sits in the heart of the rosette itself; use low power
	Emission?/ reflection?	10	2"	06392n0844	Monoceros	Fan-like shape; variable outline and internal details

Fig. 17.4. M42, the Great Nebula in Orion. (AC)

M42, the "Great Nebula in Orion," is so stunning and brilliant in virtually any telescopic view that it must surely be the most spectacular single object in the sky of the Northern, if not the Southern, Hemisphere (Fig. 17.4). The Trapezium reveals so many tiny young stars (not visible here because they are washed out by the exposure) that you may have difficulty trying to catalog them. The nebula itself is composed of swirling and twisted shapes, outspread like a great wingspan. Also note the regions of dark nebula, which obscure parts of the illuminated regions to even more dramatic effect.

NGC2261 – The extraordinary varying structure of Hubble's "Variable Nebula" seems to defy the laws of physics (Fig. 17.5). If dark shadows cast by dust or similar are responsible for shadows from its bright stellar point to be cast along its 3-light-year length, the changes would have to occur faster than the speed of light, apparently precluding such origins. Note the striking semicircle of stars surrounding the nebula at the bottom of the image.

NGC 2264 "Christmas Tree Cluster"	Open cluster			06411n0953	Monoceros	Large and spread out; much nebulosity present, including dark "Cone Nebula"
NGC 2287 (M41)	Open cluster	6	30"	06470s2044	Canis Major	
NGC 2323 (M50)	Open cluster	6	10"	07032s0820	Monoceros	
NGC 2359 "Thor's Helmet"	Emission Nebula	11	6" × 8"	07186s1312	Canis Major	Contains many stars

NGC 2371/2	Planetary	12.5	50″ × 30″	07256n2929	Gemini	Faint but interesting, because two bright zones give the impression of double ends, hence double designation
NGC 2392 "Eskimo Nebula"	Planetary	8	40″	07292n2055	Gemini	10m. central star; high powers and conventional, unfiltered viewing show it best
NGC 2419	Galaxy Sc	8.8	16″ × 10″	07369n6536	Camelopardalis	
NGC 2403 "The Intergalactic Wanderer"	Globular	11.5	2″	07381n3853	Lynx	Most distant Milky Way globular; interesting to see; image intensifiers may resolve stars
NGC 2420	Open cluster	9	7″	07385n2134	Gemini	Remote; resolution difficult
NGC 2422 (M47)	Open cluster	5	20″	07366s1430	Puppis	
NGC 2437 (M46)	Open cluster	8	25″	07418s1449	Puppis	With planetary NGC 2438
NGC 2438 (M46)	Planetary	10	65″	07418s144	Puppis	With 17m. central star
NGC 2440	Planetary	11.5	50″ × 20″	07419s1813	Puppis	Complex appearance, in some ways like a small planet Saturn, lobes; 16m. central star
NGC 2447 (M93)	Open cluster	7	18″	07446s2352	Puppis	
NGC 2477	Open cluster	7	25″	07523s3833	Puppis	Densely populated
NGC 2539	Open cluster	6.5	20″	08107s1250	Puppis	
NGC 2547	Nebula	5.5	15″	08100s4910	Vela	
NGC 2613	Galaxy Sb (edge-on)	10.9	6.4″ × 1.5″	08334s2258	Pyxis	
NGC 2682 (M67)	Open cluster	7	15′	08504n1149	Cancer	
NGC 2683	Galaxy Sb	10.6	9″ × 1.3″	08527n3325	Lynx	Almost edge-on; magnificent; fine dust lane
NGC 2808	Globular	6	7″	0912s6450	Carina	Fine object
NGC 2841	Galaxy Sb	10.3	6.2″ × 2″	09220n5058	Ursa Major	
NGC 2903	Galaxy Sb/Sc	9.7	11″ × 4.7″	09322n2130	Leo	Elongated
NGC 2964	Galaxy Sb/Sc	11.9	2.3″ × 1.1″	09429n3151	Leo	
NGC 2976	Galaxy Sc/Sd/Irr	10.8	3.4″ × 1.3″	09432n6808	Ursa Major	Mottled

| NGC 3031 (M81) | Galaxy Sb | 8.9 | 18″ × 10″ | 09556n6904 | Ursa Major | One of the most beautiful spirals in the sky; spiral structure not apparent or revealed in real time |
| NGC 3034 (M82) | Galaxy Irr. | 11.2 | 11″ × 2″ | 09558n6941 | Ursa Major | See other images, Chap. 14, Fig. 14.1 |

Fig. 17.5. NGC 2261, Hubble's "variable" nebula. (AC)

Fig. 17.6. M82. (AC)

One of the most stunning galaxies in the sky for the amateur observer, the explosive irregular galaxy M82 reveals all kinds of dramatic detail even with using relatively modest equipment (Fig. 17.6). However, as always, greater apertures reveal ever more of the mottled structure. It appears to be literally "coming apart at the seams."

NGC 3109	Galaxy Irr.	11.2	11" × 2"	10031s2609	Hydra	
NGC 3115 "Spindle Galaxy"	Galaxy E7/SO	10	4" × 1"	10052s0743	Sextans	Bright
NGC 3132 "Eight Burst Nebula"	Planetary	8.2	84" × 52"	10069s4021	Vela	Appears similar to RING NEBULA, with 10m. central, not illuminating, star
NGC 3184	Galaxy Sc	10.5	5.5" × 5.5"	10183n4125	Ursa Major	Face-on
NGC 3187	Galaxy SBc	13	1" × 0.3"	10178n2152	Leo	Faint
NGC 3190	Galaxy Sb	12	3" × 1"	10181n2150	Leo	Edge-on, tiny dust lane
NGC 3193	Galaxy EO	12	0.9" × 0.9"	10181n2150	Leo	
NGC 3242 "Eye Nebula"	Planetary	8.9	40"	10248s1838	Hydra	Startling appearance
NGC 3351 (M95)	Galaxy SBb	11	4" × 3"	10440n1142	Leo	
NGC 3368 (M96)	Galaxy Sb	10.2	6" × 4"	10468n1149	Leo	Fuzz
NGC 3372 "Keyhole Nebula"	Emission Nebula		80" × 85"	1044s5950	Carina	Magnificent sight, with numerous dark lanes crossing, much in the manner of the "Trifid Nebula." Contains famous variable star Eta Carinae
NGC 3379 (M105)	Galaxy E1	10.6	2.1" × 2"	10478n1235	Leo	
NGC 3384	Galaxy E7/Sc	11	4" × 2"	10483n1238	Leo	
NGC 3521	Galaxy Sb	10.2	6" × 4"	11058n000	Leo	Bright nucleus, elongated
NGC 3532	Emission Nebula		60"	1106s5840	Carina	Fine sight at low powers
NGC 3556	Galaxy Sc	10.8	7.8" × 1.4"	11115n5540	Ursa Major	Dust lanes – stellar nucleus; near "Owl" Nebula
NGC 3587 (M97) "The Owl Nebula"	Planetary	11	150"	11148n5501	Ursa Major	Surface brightness very low; dark features
NGC 3621	Galaxy Sc/Sd	10.6	5" × 2"	11183s3249	Hydra	Bordered by stars
NGC 3623 (M65)	Galaxy Sa/Sb	10.3	7.8" × 1.6"	11189n1305	Leo	Elongated

NGC 3627 (M66)	Galaxy Sb	9.7	8″ × 2.5″	11202n1259	Leo	
NGC 3628	Galaxy Sb	10.3	12″ × 2″	11203n1336	Leo	Fine sight; edge-on
NGC 3810	Galaxy Sc	11.5	3.6″ × 2.5″	11410n1128	Leo	
NGC 3941	Galaxy E3/SO	11.3	1.9″ × 1.1″	11529n3659	Ursa Major	
NGC 3992 (M109)	Galaxy SBb	10.9	6.4″ × 3.5″	11576n5323	Ursa Major	
NGC 4088	Galaxy Sb/Sc	11.1	4.7″ × 1.5″	11576n5323	Ursa Major	Mass to one side
NGC 4096	Galaxy Sc	11.5	4.1″ × 1.1″	12060n4729	Ursa Major	Edge-on
NGC 4111	Galaxy E7	11.6	3.4″ × 0.8″	12071n4304	Ursa Major	Edge-on, striking; bright star nearby
NGC 4125	Galaxy E5/SO	11.1	2.1″ × 1.1″	12081n6511	Draco	
NGC 4192 (M98)	Galaxy Sb	11	8.2″ × 2″	12138n1454	Coma Berenices	edge-on
NGC 4214	Galaxy Irr. or early SB	10.5	7″ × 4.5″	12156n3620	Canes Venatici	
NGC 4216	Galaxy Sb	10.9	7.2″ × 1″	12159n1309	Virgo	Thin edge-on with 2 others in field; near center of Virgo Galaxy Cluster
NGC 4217	Galaxy Sb	11.9	4″ × 1″	12158n4706	Canes Venatici	With dust lane
NGC 4244	Galaxy Sb	10.7	13″ × 1″	12175n3749	Canes Venatici	Edge-on; streak
NGC 4254 (M99)	Galaxy Sc	10.4	4.5″ × 4″	12188n1425	Coma Berenices	
NGC 4258 (M106)	Galaxy Sb	9	19.5″ × 6.5″	12190n4718	Canes Venatici	Fairly striking
NGC 4303 (M61)	Galaxy Sc	10.2	5.7″ × 5.5″	12219n0428	Virgo	Face-on
NGC 4321 (M100)	Galaxy Sc	10.4	5.2″ × 5″	12229n1547	Coma Berenices	
NGC 4361	Planetary	10.5	80″	12245s1848	Corvus	With 13m. central star
NGC 4372	Globular	8	18″	12260s7240	Musca	Mostly 12m. stars
IC 3568	Planetary	11.6	18″	12329n8233	Camelop-ardalis	
NGC 4374 (M84)	Galaxy E1	10.5	2″ × 1.8″	12251n1253	Virgo	Plus 2 additional edge-on galaxies
NGC 4382 (M85)	Galaxy E	10.5	3″ × 2″	12254n1811	Coma Berenices	
NGC 4406 (M86)	Galaxy E3	10.5	3″ × 2″	12262n1257	Virgo	Strong red spectrum; small elliptical galaxy nearby. At the center of the Virgo Cluster

NGC	Type	Mag	Size	Code	Constellation	Notes
NGC 4449	Galaxy Irr.	10.5	4.2″ × 3″	12282n4406	Canes Venatici	
NGC 4472 (M49)	Galaxy E3/E4	10.1	4″ × 3.4″	12298n0800	Virgo	One of the largest
NGC 4486 (M87)	Galaxy E1	10.1	3″× 3″	12308n1224	Virgo	Giant; famous jet not visible from my location
NGC 4490 "Cocoon Galaxy" with NGC 4485	Galaxy ScGalaxy Irr. or E	10.112.5	5″ ×2″; 1.3″ × 0.7″	12306n4138	Canes Venatici	Pear-shape
NGC 4501 (M88)	Galaxy Sb	10.5	5.7″ × 2.5″	12320n1425	Coma Berenices	
NGC 4526	Galaxy E7/SO	10.7	4″ × 1″	12340n0742	Virgo	Edge-on, between two 7m. stars
NGC 4535	Galaxy SBc	10.7	6″ × 4″	12343n0812	Virgo	S-shape
NGC 4552 (M89)	Galaxy E	11	2″ × 2″	12357n1233	Virgo	
NGC 4559	Galaxy Sc	10.5	10″ × 3″	12360n2758	Coma Berenices	
NGC 4565	Galaxy Sb	10.5	10″ × 3″	12363n2559	Coma Berenices	Most famous edge-on galaxy, exceptional, prominent dust lane
NGC 4569 (M90)	Galaxy Sb	11.1	7″ × 2.5″	12368n1310	Virgo	
NGC 4579 (M58)	Galaxy Sb	10.5	4″ × 3.5″	12377n1149	Virgo	
NGC 4590 (M68)	Globular	8	9″	12395s2645	Hydra	
NGC 4594 (M104) "Sombrero Galaxy"	Galaxy Sa/Sb	8.2	7″ × 1.5″	12400s1137	Virgo/ Corvus	Near edge-on, exceptional, dark "equatorial" lane
NGC 4621 (M59)	Galaxy E3E4	11	2″ × 1.5″	12420n1139	Virgo	
NGC 4649 (M60)	Galaxy E1E2	10	3″ × 2.5″	12437n1133	Virgo	
NGC 4631	Galaxy Sc	9.7	12.5″ × 1.2″	12421n3232	Canes Venatici	Edge-on
NGC 4656 (M9)	Galaxy Irr.	11	19.5″ × 2″	12440n3210	Canes Venatici	Bar/curved ends
NGC 4699	Galaxy Sa/Sb	10.3	3″ × 2″	12490s084	Virgo	
NGC 4725	Galaxy SBb	10.5	7.5″ × 4.8″	12504n2530	Coma Berenices	
NGC 4736 (M94)	Galaxy Sb	8.9	5″ × 3.5″	12509n4107	Canes Venatici	Very bright, no detail
NGC 4753	Galaxy Irr./E	10.6	2.8″ × 2″	12524s0112	Virgo	

NGC 4755 "Jewel Box"	Open cluster		10"	1253s6030	Crux	Adjacent to "COAL SACK" dark nebula; closely placed bright stars
"Coal Sack"	Dark Nebula				Crux	Adjacent to NGC 4755 Cluster – see "Jewel Box" Cluster
NGC 4826 (M64) "Black Eye Galaxy"	Galaxy Sa	8.6	7.5" × 3.5"	12567n2141	Coma Berenices	Structure visible; "black eye" is broad dust lane
NGC 4590 (M68)	Globular	8	9"	12368s2629	Hydra	
NGC 4945	Galaxy SBc	9.2	15" × 2.5"	13050s4920	Centaurus	Edge-on; outstanding
NGC 5005	Galaxy Sb	10.8	4.1" × 1.6"	13109n3703	Canes Venatici	Bright
NGC 5024 (M53)	Globular	8	10"	13129n1810	Coma Berenices	1" distant from fainter NGC 5053
NGC 5033	Galaxy Sb	10.3	8" × 4"	13134n3636	Canes Venatici	
NGC 5055 (M63)	Galaxy Sb	9.8	9" × 4"	13158n4202	Canes Venatici	Bright
NGC 5102	Galaxy SO	10.8	6" × 2.5"	13220s3630	Centaurus	
NGC 5128 Centaurus "A"	Galaxy SO/ pec	7.2	31" × 23"	13255s4301	Centaurus	Round, detail, central band visible, exceptional
NGC 5139 "Omega Centauri"	Globular	4	30"	13268s4729	Centaurus	Astounding; finest known
NGC 5189	Reflection Nebula		185" × 130"	13330s6550	Musca	
NGC 5194 (M51) "Whirlpool Galaxy"	Galaxy Sc	8.7	10" × 5.5"	13299n4712	Canes Venatici	Spectacular; visually bridging to NGC 5195 – 9.6m. Galaxy Pec. Spiral structure very apparent, much detail; infrared patches throughout

One of the greatest sights in the sky, the "Sombrero Galaxy" glows brilliantly, like a giant light fixture hanging in space (Fig. 17.7). Almost edge-on, it is easily seen dust belt appears to extend all around the structure, suggesting a dish, or perhaps more appropriately the "sombrero" hat shape that has become its hallmark. Absolutely unforgettable.

M51, the "Whirlpool Galaxy," made history by being the first galaxy to reveal spiral form in the nineteenth century (Fig. 17.8). The famous double galaxy (actually, two

Fig. 17.7. M104, the Sombrero Galaxy (AC).

Fig. 17.8. M51, the Whirlpool Galaxy (AC).

unlinked galaxies – the smaller one is well behind the larger) is even more remark-able in the intensified view because even the uninitiated can instantly see complete spiral form and subtle detail. The view here, while typical of many an amateur's visual impression in favorable circumstances, exhibits certain characteristics of image intensification. These show as a more refined appearance of the spiral arms, with less of the total halo luminosity than is seen in the conventional view. Note the subtle wisps in the spiral arms (faint evidence of gaseous nebulae), as well as the wide, dark trails between them, especially those appearing from the 2 o'clock and 8 o'clock positions.

NGC 5236 (M83)	Galaxy Sc	8	10" × 8"	13370s2952	Hydra	Strong emission spectrum in nucleus
NGC 5253	Galaxy E	10.8	4" × 1.5"	13399s3139	Centaurus	
NGC 5272 (M3)	Globular	6	18"	13422n2823	Canes Venatici	Beautiful, well resolved to center
NGC 5307	Planetary	12	15" × 10"	13510s5110	Centaurus	Use higher powers
NGC 5457 (M101)	Galaxy Sc	9	22" × 20"	14032n5421	Ursa Major	Face-on; spiral structure detectable with low powers
IC 4406	Planetary	11	100" × 35"	14224s4409	Lupus	
NGC 5746	Galaxy Sb	11.7	6.5" × 0.8"	14449n0157	Virgo	Edge-on, dust belt, bright condensations
NGC 5866	Galaxy SO	11.1	2.9" × 1"	15065n5546	Draco	Elongated with prominent thin dust lane; exceptional and refined
NGC 5904 (M5)	Globular	6.2	13"	15186n0205	Serpens	Superb
NGC 5907 "Splinter Galaxy"	Galaxy Sb	11	11" × 0.6"	15159n5619	Draco	Edge-on, needle-shape with some mottling and dust obscuration possible with larger apertures
NGC 5986	Globular	8	5"	15460s374	Lupus	
NGC 6026	Planetary	12.5	50"	16014s3432	Lupus	Ring
NGC 6058	Planetary	12	25" × 20"	16044n4041	Hercules	
IC 4593	Planetary	11	13" × 10"	16122n1204	Hercules	
NGC 6093 (M80)	Globular	8	7"	16170s2259	Scorpius	Small, bright; appears to radiate spikes of mostly 14m. and 15m. stars
NGC 6121 (M4)	Globular	7.4	20"	16236s2632	Scorpius	Large, appearing, open, and relatively sparser; known for striking loops and chains of brighter stars; bright "equatorial" bar of central stars
NGC 6153	Planetary	11.5	20"	16315s4015	Scorpius	
NGC 6205 (M13)	Globular	5.2	23"	16417n3628	Hercules	Exceptional; grandest in N. Hemisphere; look for "propeller" lanes. See image in Chap. 3, Fig. 3.1
NGC 6210	Planetary	9.7	20" × 16"	16445n2349	Hercules	With 12.5m. central star/ some detail; oval shape
NGC 6218 (M12)	Globular	8	10"	16472s0157	Ophiucus	Fairly sparse

NGC 6231	Open cluster	6	15″	16540s4148	Scorpius	
NGC 6254 (M10)	Globular	7	8″	16571s0406	Ophiucus	Near to globular NGC 6218 (M12)
IC 4634	Planetary	12	20″ × 10″	17016s2150	Ophiucus	17m. central star
NGC 6266 (M62)	Globular	6.5	6″	17012s3007	Scorpius	
NGC 6273 (M19)	Globular	7	6″	17026s2616	Ophiucus	Oblate; near center of Milky Way; faint star population
NGC 6302 The "Butter-fly" or "Bug Nebula"	Planetary?	9.6	2″ × 1″	17137s3706	Scorpius	Irregular shape, like a flattened figure "8"
NGC 6309	Planetary	11.5	20″ × 10″	17141s1255	Ophiucus	With 14m. central star
NGC 6326	Planetary	12	15″ × 10″	1720s5140	Ara	Use higher powers
NGC 6333 (M9)	Globular	8	4″	17192s1831	Ophiucus	
NGC 6337	Planetary	12.3	38″ × 28″	17223s3829	Scorpius	Bright circumference; inner stars
NGC 6341 (M92)	Globular	6.5	8″	17171n4308	Hercules	Uneven distribution, smaller than nearby M13, but impressive
B 72	Dark nebula		30″	17235s2338	Ophiucus	Famous S-shape, more difficult visually than B143. Use lowest power
NGC 6352	Globular	9	8″	17250s4820	Ara	Many fine stellar points
NGC 6362	Globular	8	9″	1732s6700	Ara	Stellar population similar to NGC 6362
NGC 6369 "Little Gem"	Planetary	11	28″	17293s2346	Ophiucus	Perfectly circular ring and 16m. central star, easily seen with image intensifier in my 18″ from my suburban location
NGC 6397	Globular	7	19″	17400s5340	Ara	One of the nearest globulars; majority stars 10m.
NGC 6402 (M14)	Globular	9	6″	17376s0315	Ophiucus	
NGC 6405 (M6)	Open cluster	6m	25″	17401s3213	Scorpius	Fine cluster
NGC 6475 (M7)	Open cluster	5	60″	17539s3449	Scorpius	Good visual cluster
NGC 6494 (M23)	Open cluster	7	25″	17568s1901	Sagittarius	Use lowest power; 9m. and 13m. stars
NGC 6503	Galaxy Sb	11	4.8″ × 1″	17494n7009	Draco	
NGC 6514 (M20) "Trifid Nebula"	Emission/ reflection nebula		25″	18023s2302	Sagittarius	Bright, exceptional; three dark lanes; illumi-nating star is a multiple. See other images in Chap. 14, Fig. 14.1

B 86	Dark nebula		4.5″ × 3″	18030s2753	Sagittarius	Striking near edge of cluster NGC 6520; easy to observe
NGC 6520	Open cluster	9	5″		18034s2754	Sagittarius	Enclosed by Sagittarius Star Cloud (M24); B86 dark nebula nearby, next to 7m. star, like a dark hole
NGC 6522	Globular	10.5	2″		18036s3002	Sagittarius	
NGC 6523 (M8) "Lagoon Nebula"	Emission Nebula	5	80″ × 40″		18038s2423	Sagittarius	Exceptional; with cluster NGC 6530; "Hourglass" shows prominently with image intensification

The "Trifid Nebula," M20, is actually part of the same immense gas cloud as the nearby "Lagoon Nebula," but separated by intervening darkness (Fig. 17.9). Its legendary form, cut into three distinct sections by strands of dark gas, along with its delicate pink and blue colors, makes it among the loveliest sights in the entire sky. Its brilliant multiple star, situated in the heart of the nebulosity, may be broken into its components by ever greater apertures and specialized tools; an image intensifier may show an amazing six separate stars. It is worth pointing out that the bluish portion of the nebula, at bottom left, shows less well in this image than it will in conventional viewing, because it is essentially a reflection nebula, something less easily revealed in intensified views. A bonus with an image intensifier is the resolution of its main illuminating star into no less than six components, something usually reserved for the grandest of telescopes.

Fig. 17.9. M20, the Trifid Nebula (AC).

NGC 6528	Globular	11	1″	18048s3003	Sagittarius	
NGC 6531 (M21)	Open cluster	7	10″	18046s2230	Sagittarius	In field with "Trifid Nebula"
NGC 6541	Globular	6	6″	18080s4340	Corona Australis	
NGC 6543 "Cat's Eye Nebula"	Planetary	8.6	22″ × 16″	17586n6638	Draco	Exceptional; helical structure partly resolved
NGC 6559	Nebulous fragment		5″	18068s2408	Sagittarius	Contains 10m. Star. Possibly connected to the Lagoon Nebula
NGC 6567		11.5	11″ × 7″	18137s1905	Sagittarius	15m. central star
NGC 6572	Planetary	9.5	15″ × 12″	18121n0651	Ophiucus	12m. central star; twisted appearing main core
B 92	Dark Nebula		15″ × 10″	18155s1814	Sagittarius	Prominent, near edge of Small Sagittarius Star Cloud
NGC 6603	Open cluster		4″	18184s1825	Sagittarius	Enclosed by M24 – Sagittarius Star Cloud
NGC 6611 (M16)	Open cluster	6.5	25″	18188s1347	Serpens	With "Eagle Nebula," emission, reflection and dark
NGC 6618 (M17) "Omega Nebula"	Emission	6	45″ × 35″	18208s1611	Sagittarius	Exceptional detail; embedded stars

When the Milky Way lies high in the skies many of the great nebulae are visible (Fig. 17.10). Because they lie within the galaxy's stellar arms, we will find most of them here, for they comprise the stuff of which stars are made. Notable examples can be found near the hub of the galaxy, in constellations such as Sagittarius, Scorpio, Scutum, Ophiucus, or Serpens. The "Lagoon Nebula," M8, is one of the best. Consisting of swirling clouds of gas, excited into luminosity, together with a most beautiful imbedded cluster of hot young stars, the Lagoon Nebula is one of the most dramatic of all the emission nebulae. Perhaps the most interesting aspect of the nebula is the presence of so much dark gas, superimposed on the illuminated gases, giving rise to the "lagoon" itself, among other things. Note the small bright section of illuminated gas toward the left of the image, appropriately known as the "Hourglass." This feature is particularly striking under image intensification.

There is nothing quite as startling as the sight of the bright yellowish nebula, the "Omega Nebula," M17, hanging like a great swan in a sparkling sea of jewels (Fig. 17.11). You can see the swan's feathers, and even its folded wing. The view here is quite typical of its appearance with all types of viewing; its brightness and well-defined outlines make it one of the most easily seen of all nebulae. It is instructive how dark the region in front of the "swan" appears when compared with the nebulosity surrounding the rest of it. There appear to be no other stars showing through this darker region, so it must be an illusion and cannot be caused by unlit gases or dust.

Fig. 17.10. The Lagoon Nebula, M8. (AC)

Fig. 17.11. M17, the Omega Nebula. (AC)

NGC 6626 (M28)	Globular	8	6'	18245s2452	Sagittarius	Bright and dense
NGC 6629	Planetary	10.5	15"	18257s2312	Sagittarius	Pale disc; 13.5m. central star
NGC 6637 (M69)	Globular	7.5	4'	18314s3221	Sagittarius	Unimposing in smaller telescopes; resolution of 14 and 15m. stars needs larger apertures; near 9m. star
NGC 6656 (M22)	Globular	6	18'	18364s2354	Sagittarius	Exceptional; large, open, bright and resolved

Magnificent M22 consists of a wide range of star illuminations, making it appear to have many large stars throughout its structure (Fig. 17.12). Of course, they only appear this way because of the relative brightness of these stars against the globular background of lesser stars; one's eyes thus perceive them as larger diffraction disks. Because of its grand size in the eyepiece, M22 appears to be among the greatest globulars in the sky, despite its relative loose form and modest size. From dark skies it will appear at its best, with many stars making up and filling in its periphery, giving it almost unrivalled stature. In truth, it looks pretty impressive even in city conditions, although few can match it when observed far from the bright lights of civilization.

Fig. 17.12. M22. (AC)

NGC 6681 (M70)	Globular	8	4"	18422s3218	Sagittarius	Uneven stellar distribution
NGC 6694 (M26)	Open cluster	9.5	9'	18452s0924	Scutum	
NGC 6705 (M11) "Wild Duck Cluster"	Open cluster	6	12'	18511s0616	Scutum	Impressive, dense, dark nebula nr. north
NGC 6715 (M54)	Globular	9	6"	18551s3029	Sagittarius	Compact and bright; remarkable, apparently as magnificent as Omega Centauri, although situated outside our own galaxy. Belonging to the Sagittarius Dwarf Galaxy – not to be confused with NGC 6822 – it requires larger apertures to resolve any of its stars
NGC 6720 (M57) "Ring Nebula"	Planetary	9	80" × 60"	18536n3302	Lyra	Exceptional; marvelous object; subtle detail in ring visible, as well as central star

The "Ring Nebula," M57, is one of the most eagerly viewed sights in the sky (Fig. 17.13). Aside from its stunning appearance in the eyepiece, part of its lure is the ever-present challenge of seeing the central illuminating star. The image below is generally representative of its visual appearance. In this image, made with modest equipment, not only is the central star an easy mark but all kinds of variations of

Fig. 17.13. M57, the Ring Nebula. (AC)

intensity and bright knots are very apparent. A second star near to the central star is either in front or behind the nebula and has no role in its illumination.

NGC 6726/9	Bright reflection nebulae			19017s3653/ 19019s3657	Corona Australis	Small
NGC 6744	Galaxy SBc	10.6	9″ × 9″	19090s6350	Pavo	
NGC 6751	Planetary	12	20″	19059s0600	Aquila	Faint but visible; clear defined oval shape and 13m. central star
NGC 6752	Globular	7	15″	19110s5950	Pavo	Outstanding
NGC 6779 (M56)	Globular	8	5″	19166n3011	Lyra	Consisting of mostly 11m. – 14m. stars; unusual location for a globular; challenging for smaller scopes to resolve
NGC 6781	Planetary	12.5	105″	19184n0633	Aquila	15.5m. central star
NGC 6809 (M55)	Globular	7	15″	19400s3058	Sagittarius	Large and looser; stars mostly fainter than 11m.
NGC 6818	Planetary	10	22″ × 15″	19440s1409	Sagittarius	15m. central star difficult; mottled disc framed by triangle of stars; near Galaxy NGC 6822 –11.2m.
B143	Dark Nebula	30″		19414n1101	Aquila	Celebrated irregular shape in middle of star field
NGC 6822	Galaxy Irr. dwarf	11.2	20″ × 10″	19449s1448	Sagittarius	"Barnard's Galaxy"; like small Magellan Cloud, but appears more like small, sparse cluster; planetary nebula NGC 6818 in same field
NGC 6826 "Blinking Nebula"	Planetary	8.8	25″	19448n5031	Cygnus	Round, with 11m. central star; v. good at high power
NGC 6838 (M71)	Globular?	9	6″	19538n1847	Sagitta	Rich and compact; lacks a dense core; stars approx. 12m.
NGC 6853 (M27) "Dumbbell Nebula"	Planetary	8	8″ × 5″	19596n2243	Vulpecula	Exceptional

NGC 6857	Planetary		40″	20019n3331	Cygnus	
NGC 6886	Planetary	11	9″ × 6″	20127n1959	Sagitta	16.5m central star
NGC 6864 (M75)	Globular	8	3″	20061s2155	Sagittarius	Fairly bright, compact and dense; most stars 17m.
NGC 6888 "Crescent Nebula"	Emission Nebula		18″ × 12″	20120n3821	Cygnus	Faintly visible
NGC 6891	Planetary	10	15″ × 7″	20152n1242	Delphinius	11m. central star
NGC 6905 "Blue Flash Nebula"	Planetary	12	44″ × 38″	20224n2005	Delphinius	Disc, with 14m. central star; partially framed by four prominent stars
NGC 6960/6992 "Veil Nebula"	Emission Nebula			20457n3043 /20564n3143	Cygnus	Large, lengthy filamentary structures
IC 5067 "Pelican Nebula"	Emission Nebula		80″	20469n4411	Cygnus	Possible to see in same manner as nearby NGC 7000, though it is fainter
NGC 6981	Globular	8.6	3″	20535s1232	Aquarius	15m. brightest stars
NGC 7000 "North American Nebula"	Emission Nebula		100″	20588n4420	Cygnus	Vast; difficult to see best with very low powers
NGC 7008	Planetary	12	85″ × 70″	21006n5433	Cygnus	Heart-shaped
NGC 7009 "Saturn Nebula"	Planetary	8	25″	21042s1122	Aquarius	Exceptional
NGC 7023	Reflection Nebula		18″	21005n6810	Cepheus	One of the brightest, with dark lanes
NGC 7026	Planetary	12	25″ × 16″	21063n4751	Cygnus	Appears visually like smudged elongated double spot; 15m. central star; bright star adjacent
NGC 7027	Planetary	9	18″ × 11″	21071n4214	Cygnus	Prominent star and two separate lobes on one side
NGC 7048	Planetary	11	60″ × 50″	21142n4616	Cygnus	Relatively difficult object with 18m. central star
NGC 7078 (M15)	Globular	6.5	10″	21071n4214	Pegasus	Exceptional, resolved, irregular
NGC 7089 (M2)	Globular	6	7″	21335s0049	Aquarius	Outstanding, resolved; look for dark lane N.W.
NGC 7099 (M30)	Globular	8	6″	21404s2311	Capricorn	Elliptical shaped
IC 5146 "Cocoon Nebula"	Emission Nebula		12″ × 10″	1534n4716	Cygnus	Low brightness; difficult

NGC 7293 "Helix Nebula"	Planetary	6.5	12"	22296s2048	Aquarius	Spread out, but visible in moderate apertures, along with central star
NGC 7331	Galaxy Sb	10.4	10" × 4"	22371n3425	Pegasus	Magnificent; thick dust belt on one side, some spiral detail
NGC 7354	Planetary	13	30"	22404n6117	Cepheus	16.5m. central star
NGC 7479	Galaxy Sb	10.8	4" × 3.1"	23049n1219		Curved arms

The "Dumbbell Nebula," M27, is one of the grandest planetary nebulae known (Fig. 17.14). In this view it is possible to see the complex striations that can sometimes be detected at the eyepiece. Notice the lines of small stars that seem to be within the "bubble" of the nebula itself, although they are well outside it. One star is within it, the nebula's central star, which is easy to spot. The chief difference between this intensified image and a conventional view would be the reduction in overall glow as we see here, although it gains in revealing the lines of tiny stars superimposed and subtle detail within the structure of the nebula itself.

Fig. 17.14. The Dumbbell Nebula, M27. (AC)

Fig. 17.15. NGC 7479. (AC)

The strangely twisted S-shaped barred spiral, NGC 7479, is fascinating to many (Fig. 17.15). With its unmistakable appearance, the galaxy frequently graces the pages of many astronomy books. Nothing is more surprising than to be able to easily resolve its well-known shape live at the eyepiece! Typical of many galaxies we will seek out, it will be relatively small in the low to moderate power field of view. Not bright enough to withstand too high a magnification, such small scales are something we will need to adjust to in live viewing.

IC 1470	Planetary	12	70″ × 45″	23052n6015	Cepheus	Fan-like irregular shape
NGC 7635 "Bubble Nebula"	poss. Planetary		205″ × 180″	23207n6112	Casseopeia	8m. central star
NGC 7654 (M52)	Open cluster	7	12″	23242n6135	Casseopeia	Unusually, this Improves with aperture
NGC 7662	Planetary	8.5	32″ × 28″	23259n4233	Andromeda	Barnard's celebrated nebula; exceptional, detail
NGC 7789	Open cluster	10	20″	23570n5644	Casseopeia	900 + stars
NGC 7793	Galaxy Sd	9.7	6″ × 4″	23578s3235	Sculptor	
NGC 7814	Galaxy Sa/ Sb	12	5″ × 1″	00033n1609	Pegasus	Fine and prominent equatorial dust lane divides bright core like an arc

Index

A

AAVSO, 203
Adirondack Video Astronomy, 41
Albategnius, 94
Albedo features, 131, 139
Alpha Centauri, 191
Alphonsus, 94
ALPO, 118, 140
Altazimuth mounting, 14–15, 27
Amateur Telescope Makers of Springfield, 15
Amateur Telescope Making, 5, 19, 27, 179
Aperture, 11, 16, 22, 25, 68, 84, 85, 91, 106, 109,
    133, 170, 179, 228
Apochromatic refractor, 14, 53
Apollo landing sites, 72
Apollo missions, 63, 65, 67, 71, 73
Apollo XI, 72
Apollo XV, 74
Aristarchus, 66, 97
Ashen Light, 155
Asteroids, 107, 161, 166
Astrophysics, 210
Astrovid 2000, 41, 65, 84, 172
Atlas of the Moon, Antonin Rukl, 67
A Traveler's Guide to Mars, by William K
    Hartmann, 130
AutoStar, by Meade, 174

B

B143 in Aquila, 195
B72 in Ophiuchus, 195
B86, the 'Ink Blot', 195
Balance, 29
Barlow lens, 20, 21, 36, 39, 42
Barnard, E. E, 148
Beehive, 188
Betelguese, 181
Binocular viewers, 35
Binoculars, 10, 164
Binomite solar binoculars, 173
Boyer, Charles, 154
Broadband filters, 31
Byrgius, 97

C

California Nebula NGC 1499, 195
Canals, Martian, 130
Cassini's Division, 125, 147, 149
Catadioptrics, 13, 30, 48, 53
Catalog, 229
Cat's Eye Nebula NGC 6543, 198
CCD cameras, 24
CCD chip, 41, 88
CCD chip saturation, 88
CCD images, 117
CCD imaging, 12, 24, 31, 84, 101, 186,
    188, 205
CCD planetary images, 118
CCD video, 6, 198, 205
CCD video cameras, 6, 24, 40, 43, 121, 193, 199
CCD video imaging, 209
Celestial Handbook, by Robert Burnham, 187

Celestron, 40, 212
Ceres, 162
Cheshire Eyepiece, 48, 50
Clavius, 69, 93
Cleaning optics, 52
Cloud cover, on Venus, 153–155
Cold, 56, 187
Collimation, 48, 49
Collins Electro Optics, 32, 37, 210
Color, 105, 110, 115, 128, 141, 149, 157, 167, 186,
    187, 196
Color filters, 24, 32, 69, 105, 128
Color video images, 118
Coma, 28, 54
Combining video frames and drawing, 120
Comet Halley, 163
Comet Holmes, 164
Comet McNaught, 181
Comets, 107, 163, 221
Contrast, 51, 110, 180
Cooling, 54
Copernicus, 97
Coronado Instruments, 104, 172, 174, 175
Crab Nebula M1, 196, 202
Craterlets, 66, 96
Crayford focuser, 30
Crepe ring, 114, 125, 148
Cygnus, 189, 192
Cylindrical projections, 141

D
Dark adaptation, 56
Dark nebulae, 194
Dawes limit, 146, 148
Daytime viewing, of the planets, 178
Deep space, 127, 185, 221
Deep space imaging, 205
Deimos, 140
Delphinius, 190
Diffuse nebulae, 192
Digital cameras, 24, 84, 210
Digital circles, 15, 23, 25
Dobson, John, 5, 15
Dobsonians, 13, 15, 16, 21, 28
Drawing, 84, 110, 115, 206
Dumbbell Nebula M27, 198
Dust storms, Martian, 133
Dwarf planet, 132, 161

E
Eagle Nebula M16, 193, 195
Electric focuser, 17, 30
Emission nebulae, 192
Enke Division, 148
Equatorial Mounting, 18, 27

Equatorial tracking, 26
Equatorial zone, Jupiter, 143
Erfle eyepiece, 20, 29
Eris, 162
Eros, 167
Eskimo Nebula NGC 2392, 198
Eye Nebula NGC 3242, 198
Eyepiece projection, 170
Eyepieces, 28

F
Faculae, 171
Festoons, 141
Field rotation, 27
Filters, 128, 224
Finder scope, 23, 179
Focal ratio, 19, 42, 50, 53, 85
Focusers, 30
Fork type Equatorial, 27
Frame integrated CCD video, 192
Frame integrating CCD video cameras, 42, 43, 84,
    202, 211
Frame stacking, 87
Full Moon, Michael Light, 67

G
Galaxies, 199
Galaxies, edge-on, 200
Galaxies, elliptical, 200, 201
Galaxies, face-on, 200
Galaxies, irregular, 200
Galilean satellites, 146
Ganymede, 146
German equatorial, 27
Ghost of Jupiter NGC 3242, 198
Ghosting, 106
Global climate change, 104
Globular clusters, 189
Go-to telescope, 11
Great Nebula in Orion M42, 194, 196
Great Red Spot, 141, 144, 145
Grinding and polishing machines, 19

H
Hadley Rille, 73, 76, 79, 80
H-alpha filter, 172, 175
Hatfield, Henry, 174
Hatfield's Lunar Atlas, 67
Helical focuser, 30
Hellas, 130, 136
Hercules Cluster M13, 188
Herschel, 197, 228
Homebuilt telescopes, 16, 22
Horsehead Nebula, 31
Hubble Space Telescope, 133, 147, 149, 163

Huygenian eyepiece, 20, 171
Hyginus Rille, 96

I
Ida, 167
Image intensifiers, 7, 32, 37, 39, 41, 43, 107, 163,
    172, 187, 189, 199, 200, 206, 208, 209, 212,
    228
Imaging, 24, 109, 209
Insulation, 54
Intergalactic Wanderer NGC 2419, 190
Internet, 215
IR filter, 172
ITT, 32

J
JMI, 20, 22, 30, 35
Juno, 162
Jupiter, 105, 111, 116, 120, 128, 140, 179

K
Keck Telescope, 158
Kellner eyepiece, 20, 29
Kuiper Belt, 107, 161–162

L
La Perouse, 65
Lagoon Nebula M8, 195
Large diffuse nebulae, 195
Laser collimators, 48
Lick 36-inch refractor, 148
Light baffling, 53
Light pollution, 25
Light pollution filters, 31, 42, 178
Limb hazes, 122
Lines of resolution, 41
Location, 45
Lowell, Percival, 132
Lumicon, 31
Lunar cartography, 84
Lunar filter, 37, 68
Lunar fly-by, 65
Lunar imaging, 84
Lunar jaggedness, 70, 83, 91
Lunar limb, 64, 71
Lunar maria, 96
Lunar terminator, 91

M
M1, 234
M104, 243
M13, 34, 190, 192
M17, 247
M2, 34, 191
M20, 207, 246

M22, 191, 249
M27, 252
M3, 192
M31 in Andromeda, 190
M33, 190, 201, 230
M4, 192
M42, 236
M5, 191
M51, 242
M54, 33, 190
M57, 250
M8, 247
M82, 201, 208, 238
Magellanic Clouds, 191
Maginus, 92
Magnification, 11, 29, 42, 88, 133, 147
Mare Acidalium, 114
Mare Crisium, 96
Maria, 96
Mariner 10, 153
Marius Rille, 96
Mars, 105, 113, 120, 123, 128, 129, 179
Mars, map, 138
Martian polar caps, 114, 134, 142
Martian winds, 131
Meade, 21, 172
Meade ETX-90, 174
Mercury, 106, 152, 179
Messenger spacecraft, 153
Messier, 97
Meteors, 107
Milky Way, 185, 186, 189, 190, 195,
    198, 221
Minor planets, 107, 166
Moon, 62, 216
Moretus, 86, 92
Motor drive, 17
Mount Hadley, 73, 77
Mount Hadley Delta, 73, 76, 78
Mount Palomar 200-inch, 17, 20, 33, 109
Mountings and bases, 55
Mylar film, 172

N
Nagler eyepiece, 29, 171
Nagler, Al, 28
Narrowband filters, 31, 192
NASA images, 72
NASA website, 72, 102
Nebulae, 186, 192
Neptune, 107, 157, 160
Newtonian reflector, 13, 16, 48, 52, 54
NGC 2261, 238
NGC 2440, 195
NGC 253, 201, 212, 213

NGC 2903, 201
NGC 40, 198
NGC 4565, 201
NGC 5128/Centaurus A, 201
NGC 5195, 200
NGC 5866, 33
NGC 6520, 195
NGC 7006, 190
NGC 7479, 254
NGC 891, 201, 232
Night Observer's Guide, by Kepple
    and Sanner, 187
North American Nebula NGC 7000, 195
North equatorial belt, 143, 149
Northern polar cap, 139
Novae, 202

O
Observatories, 44
Observing the Moon, Gerald North, 67
Occultations, 107
Olympus Mons, 137
Omega Centauri NGC 5139, 190, 191, 242
Omega Nebula M17, 195, 196
Open clusters, 191
Optical alignment, 47
Orion, 21, 28, 29, 174
Orthoscopic eyepiece, 20, 29

P
Paint Shop Pro 8, 177
Pallas, 162
Panoptic eyepiece, 29
Parker, Donald, 118
Perseus, 189
Petavius, 88
Phobos, 140
Photography, 85
Pickering, William and Edward, 146
Pinwheel Galaxy M33, 201
Pipe Nebula B78, 196
Pixels, 41, 85, 88
Planet X, 132
Planetary Imaging, 109, 118
Planetary nebulae, 188, 197
Planetoid, 132
Planets, 101, 105, 218
Plato, 69, 96
Pleiades Cluster, 193
Pleiades Cluster M45, 189
Pleiades M45, 189
Plössl eyepiece, 29, 32
Plutino, 161
Pluto, 132, 166
Pluton, 80
Polar alignment, 174

Polar caps on Mars, 114, 135, 139
Polar caps on Venus, 155
Polarizing filters, 153, 178
Poncet Platform, 26
Portability, 17, 20, 59
Porter, Russell, 15, 17, 19, 20
Primary mirror support, 54
Prinz, 65
Proclus, 97
Propeller Lanes, 34, 192
Ptolemaeus, 94

R
Radian eyepiece, 29
Ramsden eyepiece, 20, 171
Ray craters, 97
Recursive frame averager, 210, 211
Red shift, 132
Reflection nebulae, 193
Reflector, 12, 31
Refractor, 12, 30
Regulus, 181
Resolution, 85, 86
Rilles, 66, 96
Ring Nebula M57, 34, 198
Ritchey-Chrétien telescope, 54
Rosette Nebula NGC 2237, 195
Rosse, Lord, 34, 190

S
Sagittarius Star Cloud M24, 189
Saturn, 105, 114, 125, 128, 146
Saturn Nebula NGC 7009, 35
Saturn's rings, 114
Schröter Rille, 96
Secondary mirror, 22, 28, 48, 53
Sedna, 162
Setting circles, 17, 24
Simultaneous contrast, 134
Sky & Telescope magazine, 155
SOHO, 174
Solar limb, 178
Solar observing, 174
Solar system, 101
Solis Lacus, 130
Sombrero Galaxy M104, 200
Southern equatorial belt, 144
Southern polar cap, 139
Space sciences, 226
Spectrohelioscope, 175
Spiral formations, in galaxies, 199
Split Ring Equatorial, 13, 20
Spokes, on Saturn's rings, 117, 148, 150
St. George Crater, 76
Stability, 19
Star clusters, 189, 191

StellaCam, 172
StellaCam EX, 41
StellaCam II, 210
Stereo viewers, 25
Straight Wall, 69
Sun, 102–104, 169, 173–178, 219
Sunspots, 171, 173
Supernovae, 202, 223
Swann Hills, 76
Sword Handle NGC869/884, 189, 191
Syrtis 'blue cloud' effect, 137
Syrtis Major, 114, 130, 179

T

Tarantula Nebula, 191
TeleVue, 21, 29, 30
The Planet Jupiter, B.M. Peek, 140
Thermal equilibrium, 152, 187
Tombaugh, Clyde, 132
Tracking platform, 15
Treisnecker Rilles, 96
Trifid Nebula M20, 195, 196
Tube currents, 53
47 Tucannae, 191
Tycho, 97

U

Ultrablock filter, 31, 208
Universal digiscoping adapter, 176

Universal Digital Camera Adapter, 212
Uranus, 107, 160

V

Vales Alpes, 96
Valles Marineris, 130
Van Slyke Engineering, 30
Variable stars, 203, 221
Veil Nebula NGC 6960/6992, 192
Venera 7, 156
Venus, 106, 152, 157, 179
Vesta, 162
Video imaging, 86
Vignetting, 21, 28, 38, 211
Violet filter, 155, 156
Visual astronomer, 101
Visual Astronomy in the Suburbs, 42, 137
Visual Astronomy Under Dark Skies, 35, 44
Voyager 2, 160

W

Watson, John, 104, 170, 173
Weather, 57, 223
Web cam, 24, 84
Whirlpool Galaxy M51, 199, 200
White spots, on Saturn, 147
Wild Duck Cluster M11, 189, 191
William Optics, 40
Wind, 56

GPSR Compliance

*The European Union's (EU) General Product Safety Regulation (GPSR)
is a set of rules that requires consumer products to be safe and our
obligations to ensure this.*

*If you have any concerns about our products, you can contact us on
ProductSafety@springernature.com*

In case Publisher is established outside the EU, the EU authorized
representative is:

Springer Nature Customer Service Center GmbH
Europaplatz 3
69115 Heidelberg, Germany

Batch number: 09473857

Printed by Printforce, the Netherlands